PHILOSOPHY AND THE FOUNDATIONS OF DYNAMICS

Although now replaced by more modern theories, classical mechanics remains a core foundational element of physical theory. From its inception, the theory of dynamics has been riddled with conceptual issues and differing philosophical interpretations, and throughout its long historical development, it has shown subtle conceptual refinement. The interpretive program for the theory has also shown deep evolutionary change over time. Lawrence Sklar discusses crucial issues in the central theory from which contemporary foundational theories are derived, and shows how some core issues (the nature of force, the place of absolute reference frames) have nevertheless remained deep puzzles, despite the increasingly sophisticated understanding of the theory which has been acquired over time. His book will be of great interest to philosophers of science, philosophers in general, and physicists concerned with foundational interpretive issues in their field.

LAWRENCE SKLAR is the Carl G. Hempel and William K. Frankena Distinguished University Professor at the University of Michigan. He is the author of *Space, Time and Spacetime* (1992), *Philosophy of Physics* (1992), *Physics and Chance* (Cambridge, 1995), and *Theory and Truth* (2000).

PHILOSOPHY AND THE FOUNDATIONS OF DYNAMICS

LAWRENCE SKLAR

CAMBRIDGE
UNIVERSITY PRESS

University Printing House, Cambridge CB2 8BS, United Kingdom

One Liberty Plaza, 20th Floor, New York, NY 10006, USA

477 Williamstown Road, Port Melbourne, VIC 3207, Australia

314-321, 3rd Floor, Plot 3, Splendor Forum, Jasola District Centre, New Delhi - 110025, India

79 Anson Road, #06-04/06, Singapore 079906

Cambridge University Press is part of the University of Cambridge.

It furthers the University's mission by disseminating knowledge in the pursuit of education, learning and research at the highest international levels of excellence.

www.cambridge.org
Information on this title: www.cambridge.org/9780521888196

© Lawrence Sklar 2013

This publication is in copyright. Subject to statutory exception and to the provisions of relevant collective licensing agreements, no reproduction of any part may take place without the written permission of Cambridge University Press.

First published 2013

A catalogue record for this publication is available from the British Library

Library of Congress Cataloging in Publication data
Sklar, Lawrence.
Philosophy and the foundations of dynamics / Lawrence Sklar.
pages cm
Includes bibliographical references and index.
ISBN 978-0-521-88819-6 (hardback)
1. Dynamics. 2. Science – Philosophy. I. Title.
Q175.32.D96S55 2012
531′.1101 – dc23 2012021833

ISBN 978-0-521-88819-6 Hardback
ISBN 978-0-521-71630-7 Paperback

Cambridge University Press has no responsibility for the persistence or accuracy of URLs for external or third-party internet websites referred to in this publication, and does not guarantee that any content on such websites is, or will remain, accurate or appropriate.

For Max and Mina

Contents

CHAPTER I

Introduction

I.I THE GROWTH OF THEORIES

A very naïve view of science might go something like this: Scientists encounter a range of observable phenomena for which they have no explanatory account. Hypotheses are generated from the imaginations of the scientists who seek to explain the phenomena in question. These hypotheses are tested against the experimental results. If they fail to successfully account for those results, the hypotheses are rejected as unsatisfactory. But if they succeed in predicting and explaining that which is observed, they are accepted into the corpus of scientific belief. Then scientific attention is turned to some new domain of, as yet, unexplained phenomena.

This simple-minded picture of science has been challenged for a variety of reasons. Some are skeptical regarding the possibility of characterizing theory-independent realms of observational data against which hypotheses are to be tested. Others have noted the way in which the testing of hypothesis by data is a subtle matter indeed. It has often been noted, for example, that even our best, most widely accepted fundamental theories often survive despite the existence of "anomalies," observational results that are seemingly incompatible with the predictions of the theories.

This book is meant to challenge the naïve view as well. But the failure of the naïve view to do justice to how science really works is worth considering from a perspective that has, perhaps, not yet received the full attention it deserves. The simple view of theories is one that fails to do adequate justice to the fact that a fundamental theory can play its part in received science over a long period of time. Without making too much of a metaphor, it is useful to compare the life of a theory over time with the life of a living being. Theories have their "fetal" stage, playing a role in science even before they exist as fully formed hypotheses. One might speak of a "pre-theory" stage in the life of a theory. When theories are first fully hypothesized and first accepted into the body of scientifically accepted belief, they exist in their first "formative" state. But, just as a living being matures, and in maturing changes its aspect in deep and important ways, theories too may, over time, develop and change. Indeed, some years, decades, or even centuries, after a theory has first been accepted into the scientific corpus, it may still be playing a fundamental explanatory role in science. But what that theory looks like, indeed, what that theory *is*, in the later stages of its existence, may be something very different from what it was taken to be when it first appeared in science.

In its later stages, the account a theory gives of the phenomena it explains may have a very different form from the account it gave in its earlier stages. In fact, there may be good reason to say that in its later guises the theory may even deal with quite different phenomena than those in the domain of the theory in its earlier incarnation, or it may treat what we might call the same phenomena, but in such a way our very characterization of those phenomena may have taken on a quite novel and distinct form. Yet it will still be appropriate to speak of this later theory as being "the same theory" as the one we had years before in the theory's infancy. It is not fair to say that the earlier theory has been refuted by its newer incarnations, nor that the newer version of the theory has replaced the old. It is as appropriate to speak of one and the same theory over time, despite the remarkable transitions the theory has undergone, as it is to speak of one and the same person in infancy, adolescence, maturity and old age. Once again pushing a metaphor rather far, it is even appropriate to speak of theoretical senescence and, further, to contemplate the remains of a theory even after its demise, its corpse as it were.

Theories can be narrow in their scope and "shallow" in their place in the overall hierarchy of our theoretical description of the world. On the other hand, they can be of very broad scope, indeed, and deeply entrenched at a fundamental place in the overall scientific scheme of things. Even a broad-scope and fundamentally placed theory can occupy the esteem of science for a very short period. We shall see an example of this when we look at Descartes' dynamics and cosmology. And a narrow and shallow theory can have a long life span. But most interesting for our purposes are theories whose scope is broad, whose place in science is fundamental, and whose life span is greatly extended. It is from theories of this kind that we will learn most about the life-history and development of a theory in science.

We will focus our attention on one such theory, the "mother of all theories" in fundamental physics. The theory that will be the object of our attention is sometimes called Newtonian dynamics and sometimes pre-relativistic classical dynamics, or, more briefly, just classical dynamics.

What is classical dynamics? It is a theory that encompasses concepts designed to allow us to describe matter in motion through space during intervals of time, namely the concepts of what is called kinematics, and the concepts needed to give an explanatory account of just why matter in motion moves as it does, that is to say the concepts of what is called dynamics. It is a theory that held the throne as the ruling explanatory account of theoretical physics from the time of Newton's *Principia* in the last third of the seventeenth century until the special relativistic, general relativistic and quantum revolutions changed dynamics forever in the first third of the twentieth century. Its evolution during its reign as dominant explanatory account for over two hundred years was one of enormous richness and complexity. The many aspects of the evolving nature of this theory provide wonderful paradigm examples of almost any thesis one would wish to illustrate in a theoretical account of how theories are born, mature and age. It is by constant reference to the history and to the nature of classical dynamics as it grows and evolves that we shall learn our methodological lessons about the life of a theory.

The pre-history of classical dynamics goes back to ancient Greek astronomy and mechanics. The Greek cosmological models of the classical and Hellenistic period, culminating in the great work of Ptolemy, and the attempts to give a systematic account of change in general, and motion in particular, of Aristotle, as well as the brilliant insights into statics of Archimedes, provided the basis on which all further scientific understanding of motion and its causes was ultimately built. Underlying this early science, and also fundamental to the later development of dynamics, were the deep insights into mathematics, and especially into geometry, of the Greek mathematicians.

Deeply insightful critical comments on problems with the Aristotelian account of motion and its causes date back as early as the sixth century. Further profound illumination comes from the work of the Islamic scientists on the nature of projectile motion, and later from the deep insights of the impetus theorists of Latin Europe in the later middle ages. The true ripening of the pre-history of classical dynamics, though, begins with the Copernican revolution in astronomy and reaches its height in the dynamical insights of early modern dynamics, especially at the hands of Galileo and Huyghens.

The birth of modern classical dynamics, if one wants a single birth date for it, would likely be taken to be the appearance of Newton's great work, the *Principia.* There is one sense in which in that work we do have many of the core elements of the theory "fully formed." But, and this will be a major theme of this book, there is another sense in which it is truly impossible to see even in Newton's brilliant systematization of dynamics the true appearance of what classical dynamics was to become in its maturity. The more than three hundred years since the publication of Newton's work have seen classical dynamics explored by many of the greatest minds of modern physical theory: the three Bernoullis, Maupertuis, d'Alembert, Euler, Lagrange, Laplace, Poisson, Hamilton, Jacobi, Mach, Hertz and Poincaré, just to name a few of the very greatest. It would be the gravest error to think of the work done by these scientists of genius applying themselves to the theory of classical mechanics as merely "adding footnote" to Newton, or "filling in the details" of the theory, or, perhaps, "reformulating the theory for the purposes of convenience and practicality of application," although they did do all of these things. Rather, their work in exploring and transforming the Newtonian theory gave to that theory its ongoing developmental growth.

It will be our purpose here to look into that growth to see what methodological lessons can be learned from it by the philosopher of science.

1.2 THE FORMULATION AND REFORMULATION OF THEORIES

Fundamental theories of physics are subjects of repeated programs of reformulation. A theory may originally be proposed in the guise of a particular formal structure. Some concepts are introduced as primitive. Other concepts defined in terms of the original conceptual basis are subsequently introduced. Some laws are proposed as fundamental. From these basic laws various consequences are deduced using the apparatus of logic and various branches of pure mathematics.

Later, however, it is seen that there may be other ways of presenting the basic concepts and laws of the theory. Sometimes a reformulation of the theory is offered that can be argued to be formally "equivalent" to the original presentation of the theory. In other cases, though, the reformulation may go beyond the original version of the theory in significant ways. New concepts not obviously definable from those of the original formulation may be introduced. New laws or structural constraints may be posited that do not merely express the content of the original laws in a different manner. For example, the new laws may add significant generality to the original version of the theory. Yet they may do this in such a way that we are inclined not to speak of them as presenting a new, more general, alternative to the original theory, but, instead, as somehow "filling out" that which was already implicit in the original theory or as "completing" the task the original version of the theory had set itself.

In many cases the reformulations of a theory are motivated by "practical" considerations. The theory is designed to solve particular problems. In the case of dynamics, it is designed to allow us to predict and explain the motions of bodies of various kinds subject to various forces and constraints. But, in its original version, it may be quite difficult to apply the theory in a fruitful way to some classes or other of problem situations that ought to come within the theory's scope. Perhaps if the theory were given a variant formulation, some of these problem cases would prove more tractable. In its new presentation the theory could, perhaps, easily be applied to the cases that proved impenetrable to the original version of the theory. But, even if a reformulation is motivated initially by practical considerations of this sort, it may turn out to be the case that, once the reformulation is in hand, it is seen to have far broader implications for our understanding of the theory. A theoretical redesign originally motivated as a matter of convenience or practicality can turn out to have unintended deep, theoretically fundamental, consequences.

In other cases the desire for deeper theoretical insight may be the explicit motivator of a theoretical reformulation. Exploring the nature of the theory already in hand, the theoretical community may discover hidden within that theory structural features that remain disguised in the original formulation of the theory. A search may then ensue for a way of reformulating the theory that brings the deep, hidden structural aspects to the forefront. Or, in exploring the theory, the theoreticians may discover fundamental analogies between structures of the theory in question and structures known from other important theoretical disciplines. Such cross-theoretical structural analogies may, once again, reveal important, theoretically interesting, aspects of the theory that might not be fully explicit when the theory is expressed in its existing form. So, again, a reformulation of the theory is sought that will make these inner structural features explicit in the surface presentation of the theory. Once these structures have been brought to the surface, they may then be employed fruitfully in further developing the theory or in applying it to still more general ranges of difficult cases.

Even purely "philosophical" motivations can lie behind the desire to reformulate a theory in fundamental ways. A philosophically minded scientist may object to the common understandings of the sort of world that a theory seems to demand in order that the theory give a correct account of nature. But the scientist may very well believe, not that the theory is incorrect, but that the inferences drawn from the theory about what the world must be like can be disputed. The scientist, that is, objects not to the theory but to its "interpretation." Perhaps if the theory were reformulated in a more satisfactory way, it may then be argued, one would no longer be misled into inferring bad "metaphysical" conclusions from the surface appearance of the theory. Perhaps, indeed, we can reformulate the theory in such a way that its better, more philosophically acceptable, interpretation can now be read off from its surface features.

Just as a theory can evolve over long periods of time, remaining one and the same theory while displaying radically different guises, the interpretive issues that plague a theory show the same ability to evolve and mutate. An interpretive issue that arises when the theory is at one stage of its development might be resolved by some later reformulation of the theory. But it might, instead, reappear as a problem area for the new reformulation of the theory, mutating in its form as the theory changes. Three of the great, classic interpretive problems of classical dynamics show this co-evolution of an interpretive problem with evolution of the theory. How should we understand the role of "force" in the theory? What understandings of the nature of space and time are necessary as underpinnings for dynamical theory? What modes of explanation need be taken as fundamental for dynamical theory? These are three interpretive questions that just will not go away. As the theory changes in its formulations and reformulations, these interpretive questions change as well. But just as we can see one and the same dynamical theory throughout its manifold reformulations, so also are we vexed by one and the same set of interpretive perplexities over the centuries, even if those perplexities evolve in their formulation as the theory itself changes.

Significant reformulations of a fundamental theory often reveal to science new perspectives on the theory and the world it describes that are of great importance. What are some of the possible consequences that can result from formulating a theory in a novel manner?

For one thing, a reformulation can supply radically new insights into the kinds of explanations of the phenomena a theory can offer. It is not simply that in the reformulated version of the theory our explanations will, of course, look somewhat different from those offered by the theory in its earlier versions. Rather, the new formulation of the theory may provide us with wholly distinct *kinds* of explanation, explanations of a sort entirely unexpected if one merely looked at the theory as it was construed in its older versions. Indeed, when examined in its new guise, the theory may lead us to reconsider, philosophically and methodologically, our very ideas of what sorts of structures may appropriately be called scientific explanations. As we shall see in some detail when we look at the sorts of explanations offered by several versions of classical dynamics, deep methodological controversy can be initiated by some claim to the effect that, once the theory had been reformulated, wholly novel sorts of explanations could be seen to receive scientific legitimacy.

A radical reformulation of a theory may lead to new insights into how the "metaphysics" of the theory is to be understood. What seemed to be the necessary ontological interpretation required in order that the theory, in its earlier versions, adequately describe the world, may seem, once the theory has received its novel reformulation, to be only an optional "interpretation" of the theory. If the earlier theory seemed to demand a world of a nature that one found philosophically objectionable, the existence of the reformulated theory may be used as part of an argument to the effect that nothing of the earlier objectionable metaphysical view of the world need be taken as imposed upon us by our desire to hold to the legitimacy and adequacy of the theory's scientific account of nature. Here again classical dynamics will provide an exemplary case of how just such uses of the possibility of theoretical reformulations to support alternative metaphysical interpretations of the theory function in practice.

A reformulation of a theory may lead to the realization that the original theory had whole realms of phenomena to which it could be applied, but which escaped notice as being within the theory's purview when the only versions of the theory available were those in its earlier formal incarnations. A theory once developed to deal with some domain of observable phenomena in the world may prove to have within its scope the possibility of explanatory accounts of many things that go on in the world that were never suspected, at the time the theory was originally devised, to be treatable by the theory's methods. We shall see just such an extension of a theory's applicability, in part suggested and motivated by the opportune systematic reformulation of the theory, once more in an example chosen from the history of the development of classical dynamics.

A theory may have as a derivative consequence of its basic assumptions some results whose generality and importance goes far beyond anything apparent from the place of these consequences in the original version of the theory. That is, there may be something that follows from a theory, but seems initially to be but one consequence of the theory among many, and whose importance may be masked by the way in which the original theory is formulated and the way in which the consequence functions within that original formalization. A reformulation of the theory may serve to reveal, finally, the true generality and importance of the consequence of the theory, the importance that had remained hidden. Indeed, once reformulated the theory may even reveal to us that these consequences of hidden importance had a generality and profoundness of such scope and depth that they would prove applicable far outside the limits of the concerns of the original theory. And the fundamental nature of these consequences might prove to be such that even when the original theory, in all its versions and guises, became rejected as no longer a true account of the world, the old theory's consequences, now in their true representation as fundamental principles, might even survive the wreckage of the theory from which they were first derived. We shall also see an example of just this process in our exploration of classical dynamics.

Accepted scientific theories are usually only transient place-holders in our scientific esteem. Even the most widely accepted and most fundamental theories must always contend with the prospect that at some future date they will be replaced by an incompatible successor theory. But scientific theory change, at least at the level of foundational physics, is rarely a kind of change in which the newer theory is wholly unrelated to the theory whose place it is usurping. In even the most revolutionary changes in foundational physics, the successor theories borrow greatly, in terms of concepts and formalism, from their predecessors. How should we react if a theory is found wanting? It might be found to be unsuccessful in correctly predicting the observational and experimental facts. Or it might be found to suffer from some internal incoherence. Or it may be the case that we must reject the theory because of its incompatibility with some other accepted theories of science that we cannot think of rejecting at the time in question.

When a theory is found to suffer from one or more of these faults, something must be done. But what is to be done may itself be suggested to us by the structure of the theory found wanting. We may, for example, be inclined to localize the failure of the existing theory in a particular one of its many components, suggesting that we might leave the other components alone in making our change to a newer theory. But, once again, how to evaluate the impact of some failure or another of a theory, and how to move ahead with changing our theory when failure is encountered, may be matters highly dependent on just how the theory found to be at fault is being formulated. Alternative reformulations of the theory may suggest quite different programs for modification and revision in our search for an improved successor to our, now to be rejected, current theory.

When successor theories are created, they often are discovered by taking apart the components of the theories they are to replace and making systematic changes in those parts of the existing theory that are to be replaced. But, then, how the existing theory is formulated, how its basic concepts and assumptions are characterized, will be highly influential in suggesting just what parts of the older theory are to be used, in modified form, to construct its new replacement, and just how these aspects of the older theory are to be changed to construct the new one. Classical dynamics was used in just such a way to construct its three famous successor theories: the special theory of relativity, the general theory of relativity and quantum mechanics. The existence of multiple reformulations of classical dynamics was invaluable in providing a rich source of suggested novel theoretical elements with which to build the theories that would replace classical dynamics itself.

1.3 THE STRUCTURE OF THIS BOOK

This book blends together a sketch of one thread in the history of science, an informal exposition of a number of aspects of one branch of theoretical science, and an attempt to derive a number of methodological conclusions in the philosophy of science from the historical and scientific material.

Let me first make some disclaimers. The history of science presented here is all quite derivative. Although I have tried to make use of original sources (at least in translation), most of the history outlined here is familiar from the established secondary sources – few as they are in the history of dynamics. Let me add that the kind of history of science with which we will be concerned is purely of the "internal" variety, and even then far from what would be expected in a work primarily devoted to the history itself. We shall look at how one idea led to another within physical science. Very rarely indeed will we touch on anything having to do with the more general historical, social or cultural context in which the science appeared. Nor will we be concerned with biographical or psychological aspects of the scientists involved. Even from the purely "internal" perspective, our focus will not be on the details of origin and influence. The history will be present, rather, as a source of illustrative material from which we might hope to extract some methodological insights.

The science presented here is familiar material, and it has been dealt with abundantly in many sources at many levels of technical sophistication. Once again, our attention to the scientific material will be selective, picking and choosing those elements in the history of classical dynamics that our useful for our methodological inquiry. Formal exposition of the science, especially the details of how results are derived and proved, will be kept to a minimum. I will delve into details only when they illuminate methodological points. The focus of the book at all times will be neither the history of classical dynamics, nor the contents of the theory itself, but, rather, how the history and internal structure of this theory can illustrate the philosophical and methodological themes outlined above.

Although this book is not a work in the history of science, I shall use a chronological history of classical dynamics as the framework on which to hang the methodological points. Overall, the book is structured temporally, starting with the pre-history of classical dynamics, and following its evolution through the Scientific Revolution, then through the period of its great development in the eighteenth and nineteenth centuries, and on through some developments in the theory, and in the theory's relation to its successor theories, in the twentieth century. Strict chronology, however, will not be adhered to, since in some cases following a single topic over a long time span will provide a more coherent organizational structure than would coming back to that topic again and again as its development occurred in time.

We will begin, then, with a sketch of the pre-history of classical dynamics. First there will be an outline of ancient Greek astronomy and ancient Greek work on the theory of motion. This will be followed by an even briefer look at the contributions made to dynamics in the period following Aristotle and preceding the great discoveries in astronomy and mechanics of the Scientific Revolution. This will include a look at an early critique of Aristotle, a glance at the work on motion of the Islamic school and a quick look at the impetus theorists of the later middle ages in Latin Europe.

Next there will be somewhat more detailed attention paid to the revolutionary results in astronomy from Copernicus to Kepler, and to the explosive developments that led to the modern science of motion in the work of Galileo, Descartes, Huyghens, Leibniz and others whose work directly impacted on that of Newton. After that we shall outline the most crucial aspects of the great Newtonian synthesis that set the stage for all further work in classical dynamics.

The work of Newton by itself led to a number of fundamental philosophical and methodological debates. The theory gave rise to issues of metaphysics: What must the world be like in order that the Newtonian theory could describe it? It stirred up issues of epistemology: What sort of theory of knowledge and inference could justify accepting the Newtonian claims? And it restored vigor to arguments about the nature of scientific explanation: What does the Newtonian theory tell us about what a scientific explanation should be like? Indeed, the philosophical controversy aroused by reflection on the Newtonian theory provided the paradigm for all future philosophical debates rising out of major revolutionary changes in foundational science. These debates will next take up our attention.

The core focus of the book, however, is on the development of classical dynamics following Newton's great synthetic work. We will be looking at how classical dynamics grew and evolved in the three centuries following the publication of the *Principia*. Here we will be concentrating for the most part on the many formulations and reformulations of the fundamentals of the theory, and how each reformulation brought with it its own issues of interest to the philosopher and the methodologist.

One of the topics to be explored will be the role of extremal principles in offering alternative fundamental dynamical laws for the theory and how such principles themselves evolved over the years. Another topic will be the difficulties encountered in applying the dynamical rules to ever more general classes of material bodies, going from the dynamics of point particles to that of rigid bodies, and finally making the theory applicable to fluids as well. Another topic to be explored is the way in which dynamics was reformulated to deal with the motion of objects subject to constraints. Whereas modifying the theory to allow it to effectively deal with the dynamics of bodies whose motion is restricted by specified constraints seems initially to be a practical problem of little fundamental theoretical interest, we shall see how the means devised to deal with the problem of constrained motion eventually opened up new conceptual vistas to the theory.

The issue of the origin, development and perfection of conservation principles is another topic we will look at. Principles of the "conservation of motion" were invoked very early in the development of the theory prior to Newton. But their role in the theory was one of constant refinement and constant re-evaluation. We shall see how they went from initial limited basic postulates, to derived consequences of other fundamental posits, to, ultimately, principles whose foundation was to be sought in symmetry considerations and whose scope outran that of the classical dynamical theory itself. After that we shall look at Hamilton's derivation of a novel set of foundational dynamical equations for the theory and how these invoked new concepts, generalized momentum and phase space, which, once again, recast the theory in such a novel framework as to open up wholly new insights into the theory and the world it described.

Hamilton was also responsible for reinvoking analogies between optical phenomena and dynamical phenomena in a way that put classical dynamics into an additional novel framework and perspective. We shall look at this.

Quite a different program for reconstruing classical dynamics was constituted by those programs, beginning in the nineteenth century at the hands of Mach, Hertz and others, whose motivation was more "philosophical" than practical. Here the desire was to enter once more into the kind of metaphysical and epistemological debates that surrounded Newton's theory from its very beginnings. Could ways be found to formally redevelop the theory so that alleged conceptual difficulties with the standard accounts of classical dynamics could finally be dealt with, whether within the science itself or by a careful interpretive program directed to it?

The late nineteenth century also saw brilliantly original work that, again, put the theory into quite a different light. This work began, as did so much other work with fundamental consequences, with practical issues. In this case the issue was the attempt to deal with long-time stability problems in dynamics and the failures of standard perturbation theory to solve those problems. This work led to Poincaré's invention of qualitative dynamics, a field that, after some lapse of time, flourished and, once more, eventuated in wholly new ways of looking at classical dynamics and construing its explanatory ability to account for the world. One branch of this new way of doing dynamics was the discovery of the radical instabilities of motion described by chaos theory, and the consequences that this branch of the theory had for our views about issues of predictability and determinism in a world ruled by the laws of classical dynamics.

We shall also explore the way in which a theory, once displaced by newer theories in the canon of accepted science, can be re-understood in a deeper way by retroactively applying concepts developed to understand the newer theories to interpretations of the older, now displaced theory. Here we shall explore how the notion of spacetime, which was invented to understand the special theory of relativity, provided new insights into the old problem of the absolute reference frame posited by Newton as essential to understand his theory. And we shall see how the notion of a curved spacetime, which was invented by Einstein to provide a relativistically acceptable theory of gravitation, could also be retroactively applied to provide deep insights into what was essential to understand aspects of gravity known to Galileo and Newton.

We shall also explore some current interpretive work on dynamics that is ongoing. One such interpretive program hopes to use the resources of modern formal logic and the philosophical understanding of the nature of theories to come to grips with the old issues of the nature of "mass" and "force" in the Newtonian theory. The other program carries forward the attempt to reconcile a relationist understanding of the nature of space and time, as first posited by Descartes and Leibniz and developed in a new way by Mach, with the existence of "absolute acceleration" used by Newton in his famous defense of "absolute space" and "absolute time."

CHAPTER 2

The pre-history of classical dynamics

2.1 SOME GREEK KNOWLEDGE AND SPECULATION

The motion of the heavenly bodies, observed as lights in the sky, provides us with a remarkable spectacle of a phenomenon describable in a small number of terms and exhibiting an easily noted regularity in space and time. This spectacle caught the attention of many cultures in the beginnings of their attempts to characterize the world as a place of some describable order. Most of the cultures were unable to get beyond the ability to discover numerical formulae that allowed one to predict recurrences in the domain of the heavens, sometimes with astonishing accuracy. In ancient Greece, however, astronomy took science further. In particular, Greek astronomy involved deep connections with Greek attempts to construct a general dynamical theory of motion and its causes. This close connection between dynamics and astronomy persisted throughout the history of classical dynamics, as we shall see. It is necessary for us, therefore, to say a little bit about some of the aspects of Greek astronomy that impinged upon Greek theories of motion.

By the time of the Greek classical era, many important facts were well known and widely agreed to. Whereas early Greek speculation about the shape of the Earth thought of it as flat, perhaps a disk of land surrounded by a circumventing ocean, it was soon an accepted fact that the Earth had the form of a sphere. Observations on how the elevations of stars changed as one moved north or south, how ships disappeared a little at a time over the horizon and how the length of a day varied with latitude could be explained only by invoking such a shape for the Earth. By analogy, models of the Moon and Sun as illuminated disks were soon replaced by accounts of these heavenly bodies as also spherical. (It is remarkable how spherical the Moon *looks*, in fact, when seen during a total eclipse.)

An extremely accurate estimate of the size of the Earth was obtained by Eratosthenes. On a day in which the Sun casts no shadow of a stick at a location in Egypt, one can measure the length of a shadow cast by a stick in Greece. A little thought about the angles involved and a fairly good idea of the distance between the sticks gives one an estimate of the circumference of the Earth. Curiously, later Greek estimates of this quantity were far inferior, taking the Earth's circumference to be far smaller than its actual value, and, eventually, leading Columbus on his hare-brained scheme to get to Japan by sailing westward.

Thinking of the Moon as a sphere, and pondering deeply the changes in the Moon's illumination over a month as it goes through its phases, the conclusion is arrived at that the Moon shines by means of light from the Sun that the Moon has reflected. It is then realized quickly that both lunar and solar eclipses have simple explanations: The former being the shadowing of the Moon from the Sun's light by the Earth and the latter an interposition of the Moon between the Sun and the Earth blocking the Sun's light.

What is most remarkable was the way in which these facts, taken together, could give rise to estimates of the distances of the Moon and the Sun from the Earth. The Hellenistic astronomers, relying on the time of passage of the Moon through the Earth's shadow during a lunar eclipse, achieved the remarkably accurate value of 59:1 for the ratio of the distance from the Earth to the Moon to the radius of the Earth. Attempts were also made to try to determine the relative distances of the Sun and Moon from the Earth. Some of these methods also relied on observations derived from lunar eclipses. Other more direct methods relied on the fact that when the moon was at half phase the triangle made of Sun, Moon, and Earth was a right triangle at the moon and on measurements on Earth of the angle between the line to the Moon and that to the Sun. The vastly greater distance to the Sun than to the Moon made these measurements difficult to perform accurately, and resulted in underestimates of the true magnitude of the Sun's distance. Although the Greeks were aware that the planets and stars were further away than the Moon and the Sun (in the case of the stars the very fact that they could be occluded by the Moon made it clear that they were behind it), speculations about just how far away they were remained just that.

But what did the overall structure of the cosmos look like? One universally prevalent idea was that the stars were rigidly attached to a sphere. The fact that they shared a common observed motion but remained in fixed place relative to one another easily led to that idea. Early cosmologies generally had the Earth at the center of this sphere.

But what of those objects whose place in the heavens changed relative to the place of the "fixed" stars? That is, where did the Moon, Sun and planets fit into the cosmic framework? Primitive ideas had the cosmic lights going under the Earth when they weren't visible. But, once we take the Earth to be a sphere, the idea that everything goes round it becomes the most plausible hypothesis.

The stars rise in the east and set in the west, suggesting a rotation of the sphere of fixed stars at a constant rate of spin. But, since the time, as judged by noting the position of the Sun, at which stars rise and set changes over the year, so that at different times of the year quite different stars are seen at night, one clearly needs to account for the fact that the position of the Sun relative to the fixed stars changes in a cyclic manner over the course of one year. The Moon also is seen to have a position relative to the fixed stars that changes over time. And there appear those almost star-like lights, the planets, or "wanderers," whose position is also variable with respect to the background of fixed stars.

The Sun moves at a constant rate from west to east relative to the fixed stars. Its full cycle takes a year, a year marked out on Earth by the clear changes in climate from season to season, changes accompanied by a length of day (from sunrise to sunset) that is maximal in mid-summer, minimal in mid-winter and equal to the length of night at the spring and autumn equinoxes. The path the Sun travels through the fixed stars is called the ecliptic. Relying on data from earlier Babylonian astronomers, the Greeks also became aware of the precession of the equinoxes, that is of the long-time cycle (about twenty-six thousand years) in which the position of the Sun at the equinox relative to the fixed stars also changes and makes its complete circuit over the millennial period.

The motion of the moon was realized to be quite complex. For one thing, its closeness to the Earth made accurate determination of its position relative to the Sun and fixed stars possible. The Moon showed a monthly cycle in which it would go from alignment in the direction of the Sun ("conjunction"), though various positions away from the Sun's direction, to a return to the direction of the Sun. These positions of the Moon relative to the Sun resulted in its phases: new moon when in the Sun's direction, full moon when in "opposition" to the Sun, and half-moon when at the intermediate quarters. The Moon can be located above or below the ecliptic line traced by the Sun. It is only when the Moon is in opposition or conjunction and on the line of the ecliptic that lunar or solar eclipses can occur.

The planets can be distinguished from the stars in that their brightness is variable, they are generally brighter than most stars, and they don't twinkle. Most important is their wandering in position relative to the fixed stars. Several features of that motion were seen to be important by the Greek astronomers. The planets generally move from west to east against the background of stars. Two of them, Mercury and Venus, are never seen more than a limited number of degrees away from the Sun. The others, Mars, Jupiter and Saturn, can have any angular position relative to the Sun. The planets move in circles against the stars that are within a bounded region containing the ecliptic (the zodiacal band) but can be above or below the Sun's path. Strikingly, the planets also show retrograde motion. That is, at some times the planets move from east to west against the background of the fixed stars rather than, as usual, from west to east.

What sort of structure of the Universe would account for this complex of observable changes in relative positions?

The model that immediately springs to mind, and the model that was the "orthodox" one in Greek astronomy for some time, had the sphere of the Earth at the center of the Universe. Around it was the rotating sphere of fixed stars. Within the sphere of the fixed stars other spheres carried the heavenly objects whose position changed relative to the fixed stars. We will shortly look at a few details of such a model.

But this model with a static Earth surrounded by moving heavenly bodies was not the only picture of the cosmos proposed. The Pythagoreans, whose writings have mostly vanished and whose thought is only obscurely transmitted to us by others, seemed to have a cosmological model in which everything, including the Earth, moved about some "central fire." In espousing heliocentrism much later, Copernicus thought he was reviving Pythagorean ideas.

Much closer to the later Copernican model were clearer suggestions by other Greek astronomers. Herakleides clearly proposed that the apparent daily rotations were the result not of a spinning of the heavens but of the rotation of the Earth around a fixed axis. He also apparently proposed that at least the inferior planets (Mercury and Venus) traveled in their circular motions about the Sun, an idea suggested by the fact that the angular distance from the Sun was constrained. It is conceivable, but not clear from the existing documents, that at least some Greek astronomers proposed the system later espoused by Tycho Brahe in which all of the planets moved in orbits about the Sun while the Moon and Sun went around a stationary Earth.

Aristarchus, and, later, Seleukos, seem to have accepted the rotation of the Earth, but also proposed that the Earth moved in a yearly orbit about a stationary Sun, a true precursor of the Copernican system. Suffice it to say that the idea that the apparent motions of the heavenly bodies were, at least in part, the illusory result of the real motion of the Earth was certainly not unknown, and indeed was widely discussed, in the Greek astronomy of the classical period.

But the motion of the Earth was generally rejected in cosmological models. Why was this so? It was not out of mere prejudice that the Earth's motion was rejected. If the Earth was spinning rapidly on its axis, as would be needed to account for the diurnal motions, why were birds and clouds not left behind? Why didn't an object tossed in the air come down some distance to the west of where it was thrown up? Why weren't objects flung off the Earth like clay from a spinning potter's wheel? If the Earth were not at the center of the sphere of the fixed stars, but was moving in a circle about the Sun at the center, why didn't the angular position of the stars vary as the year went on? Such a parallax effect could be expected, but was not observed.

2.2 THE GROWTH OF MATHEMATICAL ASTRONOMY

It is one thing to give a qualitative characterization of some proposed structure for the cosmos. It is quite another matter to construct a detailed account of the "real" motions of the heavenly bodies that would accurately account for the changing positions of these objects relative to one another as observed from the Earth. Such a program would have two distinguishable aspects. First one must construct a model of the motion of the heavenly objects that would give rise to the observed changes in the heavens. Next one must construct a physical account that explains why the heavenly objects have the motions attributed to them in the "mathematical" account.

Perhaps the most important reason why the Greeks were able to go far beyond any other culture in their construction of models of the structure and motions of the Universe was their mastery of geometry. The origins of systematic Greek geometry, and especially of the idea that geometric truths were to be established by proofs, that is by logical derivation of the hypotheses in question from "self-evident first principles," is obscure. But by the classical period, the time in which Greek mathematical astronomy properly so called began, geometry as would be recognized today in high-school textbooks was well established. Greek astronomy presents from the classical period through the great work of hellenistic Alexandria the first clear demonstration of how fundamental physical science develops out of equal parts of careful observation and mathematical systematization.

For many years the preferred model was that introduced by Eudoxus, adopted by Aristotle and improved upon by Callippus. There is an outer sphere upon which the fixed stars are mounted and which turns at a uniform rate about an axis that intersects the sphere at the celestial poles, i.e. at those points at which when observed from the Earth we see no stellar motion and about which all the stars move in circular paths. Inside of the stellar sphere are nested other spheres. Each has its axis fixed to the sphere just outside it at two points, not generally the axial points of the outer sphere, and turns about that axis with a uniform rate. Thought of as physical objects, the spheres must be transparent, or "crystalline," since we see through them to the objects mounted on each sphere up to the outermost. On some of the inner spheres the wandering heavenly bodies, Moon, Sun and planets are attached. Others serve just as intermediary devices to get the motion of the body-carrying spheres correct. All of the spheres are centered on the center of the Earth, and all turn at fixed angular rates.

The model can mathematically do a fair job on getting the observed motions right, but only a fair job. One can generate retrograde motions for the planets within the model, for example, but never, even with ever more complex layers of spheres, get that retrograde motion with all the observed angular changes in time and changes in latitude relative to the ecliptic correct.

The model of Eudoxus was taken up by Aristotle, who combined its attempt at getting the mathematical part of astronomy right with a physical account of the causes behind the heavenly motion. Let us forego discussing Aristotle's physics of the spheres until we can outline his overall dynamical scheme.

Several defects in the general principles of the Eudoxian, or homocentric, model became clear to later Greek astronomers. The planets vary in their brightness, suggesting that their distances from the Earth vary over time. But any homocentric model requires that they maintain constant distances from the Earth. The Eudoxian model requires a constant speed of rotation about an axis that goes through the center of the Earth. For the Sun Eudoxus assumes uniform angular motion about the Earth. But Medon and Euktemon had observed some decades before Eudoxus proposed his model that the lengths of the seasons were unequal. That is, the time periods from winter solstice to spring equinox, from the latter to the summer solstice, from that solstice to the vernal equinox, and from the vernal equinox back to the winter solstice were not all the same. The two fundamental principles of homocentrism, namely that the Earth was centered in the orbits of the wandering bodies and that the bodies traveled those orbits at a constant rate when observed from the Earth, must then be rejected.

Around 230 BC Appolonius introduced the epicycle into mathematical astronomy. Let the heavenly body move uniformly in a small circle whose center itself moves uniformly in a larger circle centered on the Earth. This device could serve both to represent the motion around the Earth of an object like the Sun or Moon whose angular velocity was not constant as seen from the Earth, and could serve to allow the construction of retrograde motion for the planets as well. Appolonius also realized that an equivalent construction would have the body move uniformly in a large circle whose center moved in a smaller circle about the Earth. In Appolonius we have the initiation of later Greek mathematical astronomy that seeks the most accurate mathematical representations of the observed motions of the bodies possible, but eschews physical speculation about "what is really out there" and about the causes of the motions.

A century later Hipparchus continued the development of mathematical astronomy. Hipparchus discovered the precession of the equinoxes. He also offered two models of the Sun's motion that accounted for the unequal lengths of the seasons. One invoked an epicycle. The other simply displaced the Earth from the center of the circular, uniform orbit of the Sun. The center of the orbit became known as the deferent point. Similar devices accounted for some of the known irregularity in the Moon's motion, but Hipparchus never constructed models for the motions of the planets.

Finally, in 140 AD the great mathematical astronomer and cartographer Ptolemy produced the final product of later Greek mathematical astronomy. Ptolemy added to the observational knowledge of the changing positions of the heavenly bodies. More importantly, he offered a comprehensive mathematical model that could generate to a high degree of accuracy the motions of the Sun, Moon, planets and stars.

For the motion of the Sun, Ptolemy simply adopted Hipparchus' notion of the Sun moving in a circle, the deferent circle, whose center was displaced from the Earth. Ptolemy's work on the Moon is quite interesting. Hipparchus had used an epicycle moving on a circle centered on the Earth. Ptolemy discovered some, but not all, of the additional corrections needed to get the Moon's position more accurately. Ptolemy had the Moon moving in an epicycle centered on a circle whose center was displaced from the Earth, an eccentric or deferent circle. But he had the center of the epicycle moving with a uniform angular rate relative to the Earth. Even stranger, he had the center of the deferent circle of the Moon itself taking a circular path around the Earth. The model gave much better predictions for the angular position of the Moon than did simpler models. But it had the peculiarity of forcing the Moon to have its distance from the Earth change over time to a degree that would be readily apparent in changes in the Moon's visual size, changes that were never observed. Such an "unrealistic" model was one of the motivations behind the frequent claim that the astronomer was not seeking a description of "how things really were" but only trying to find a mathematical formalism that would "save the phenomena," in this case the angular place of the Moon relative to the Sun and stars.

Ptolemy's most famous accomplishment was in his theory of the motion of the planets. The planets, first of all, move on epicycles whose centers move along a deferent circle centered at a point away from the Earth. The use of the epicycle easily explains retrograde motion, for, in some of its positions of the epicycle, the planet will be moving relative to the Earth along the epicycle path in the opposite direction from the way in which the center of the epicycle is moving along the deferent path. The retrograde motions of Mars, Jupiter and Saturn always occur when the planet is in the opposite direction from the Earth than is the Sun. To force this, all one need do is demand that the line from the center of the planet's epicycle to the planet itself is always in the same direction as the Sun is from the Earth. Actually, in a way that led later on to many difficulties, it was the *mean* direction of the Sun, that is the direction the Sun would have if it moved uniformly about the Earth, that Ptolemy employed, not the Sun's actual direction from the Earth at a particular time.

For Mercury and Venus, a similar deferent–epicycle system was used. But in the case of those planets it was the center of the planet's epicycle whose direction from the Earth was always in the *same* direction as the mean Sun. The reason for the difference in the treatment of the two classes of planets is clear if one looks back at Ptolemy's system from the modern point of view. The apparent motion of a planet is the compound result of the planet's own motion around the Sun and the Earth's motion around the Sun. This leads to two circles, a larger one and a smaller one. Ptolemy always took the larger to be the deferent and the smaller the epicycle. So, for Mars, Jupiter and Saturn, the deferent circle represents the planet's orbit, and the epicycle the projection onto the planet of the Earth's orbit. For Mercury and Venus, however, the situation is reversed, with the deferent circle as the projection of the Earth's motion and the epicycle capturing the planet's real orbital motion.

The Ptolemaic system is yet more complicated. The retrograde motions vary in size. This is easily understood since the Earth is not at the center of the deferent circle, so the angle subtended by the epicycle at the Earth varies in size. This gives an estimate of the relative distance from circle center to Earth. But if one then assumed that the center of the epicycle of the planet moved with uniform angular velocity about the center of the deferent circle, one got poor results. Ptolemy discovered that he could improve the results by having the epicycle center move with uniform angular velocity around a point on the other side of the center of the deferent circle from the Earth and just as far away from the center as the Earth was, the so called equant point. Even for Ptolemy, the device of the equant, which had been discovered by trial and error, seemed artificial to a degree not true of the remaining parts of the planetary mechanism. For Mercury even more complications were necessary; a "crank mechanism" somewhat like that introduced for the Moon was required.

More complications were required to deal with the fact that the planets had not only changing longitudes and changing angular positions relative to the Sun and fixed stars, but changing latitudes as well. That is, they sometimes appeared above and sometimes below the ecliptic, the path traced by the Sun against the stars. This required an additional circle perpendicular to the plane of the deferent circle in which the epicycle center moved in the case of Mars, Jupiter and Saturn, and an even more complex device in the case of Mercury and Venus, which had the epicycle plane tilted with respect to the plane of the deferent circle and "wobbling" over time, to get the changing latitudes correct in their correlation with the longitudes.

Ptolemy, following Appolonius, was aware that alternative mechanisms could be proposed that would give demonstrably equivalent observational results. That fact, along with the fact that some models adequate for one purpose wholly failed in other respects, like the "crank" model for the motion of the Moon, once again suggested the methodological attitude toward mathematical astronomy that anticipated much later positivism in its proposal that theories be taken not as representing some hidden reality, but as devices for making accurate predictions of the observable phenomena.

2.3 GREEK DYNAMICAL THEORY

The Greek discoveries that survive intact in current mechanics are the discoveries in statics made by Archimedes about the principles governing simple machines such as the lever and the invention, also by Archimedes, of the mathematical methods that eventually led to the calculus. One might also attribute to the atomists Leucippus and Democritus the idea that things stay in motion in straight lines unless interfered with. At least, that is what they said about the primitive atoms of their cosmology.

But it is the work of Aristotle that, one way or another, had the most important influence on the development of dynamics. Aristotle offered the only comprehensive theory of motion and its causes in the ancient Greek world. And it is Aristotle's views, often as the great errors that had to be refuted, that are the starting point for all future work in dynamics.

Aristotle's doctrine is complex, but systematic and integrated. Elements of metaphysics, cosmology and physical science are all brought into play. Aristotle's cosmology is a version of that of Eudoxus, with a stationary spherical Earth surrounded by nested heavenly spheres in uniform rotation. Aristotle is well aware of cosmologies that have the Earth in rotary motion or traveling in an orbit, and offers the arguments noted above against the hypothesis of a moving Earth or an Earth not centered in the Universe. The stationary Earth provides a reference standard against which all motions can be judged as being either true states of rest or true states of motion of some magnitude in some direction.

Aristotle views the world as composed of five basic substances, earth, water, air, fire and that fifth stuff, the "quintessence" of which the heavenly bodies are composed. While the Earth is a domain of transience and change and corruption, the heavens are a place of eternal regularity and constant, uniform, cyclic motion of unchanging beings.

Change consists in substance that remains unchanged while its forms come and go. A potter's clay, for example, remains the same stuff although its shape goes from lumpiness to pot-shaped. Forms can exist potentially in something (like the pot shape in the clay before the potter shapes it) and in actuality (like the pot shape in the clay after the pot has been formed). To account for change one must specify the stuff changed (the material cause), and the forms induced in the change (the formal cause). But one must also specify the "efficient cause" that brought about the change (the potter's hands, for example), and a "final cause," an end or purpose of the change, as well. For Aristotle even changes in nature not wrought by obvious intending beings are accountable for in final causal as well as efficient causal terms.

Aristotle tries to understand what "place" and "time" are, coming to the conclusion that the place of a thing is the enveloping stuff that bounds it and that time is a "calculable measure of motion." But time is not identifiable with any specific motion itself. But what accounts for motion?

Some motions for Aristotle are "natural," and these are contrasted with "forced" motions. The fall of an object toward the Earth is natural. There is a natural place for elements in the Universe. The natural place for things made of earth is the center of the Earth. If they are displaced from that location, it is only natural that they tend to return, of themselves, to their rightful place. Similarly fire rises, for the natural places of things are in the sequence earth, water, air, fire and quintessence, going outward from the center of the Universe to its heavenly outer reaches.

The motion of the heavenly bodies is natural as well. These motions do have an efficient cause, in that a "Prime Mover" turns the sphere of the fixed stars which then, somehow, transmits motion to the inner heavenly spheres. But the motions have a final cause, in that they are the most perfect – purely circular, constant, eternal – motions possible in a finite Universe. There are places where Aristotle thinks of the heavenly bodies as having "spirits" that in some sense guide their motion, almost as an intentional activity. Remember that in all this talk of perfect, uniform circular motion, Aristotle is working long before the later Greek astronomers took such care to do justice to all the "imperfections" they had discovered in the heavenly motions.

But, for our purposes, it is Aristotle's account of "forced" motion that is the most important aspect of his science for us to look at. A typical forced motion is that which occurs when, for example, we push a piece of furniture along the floor, moving it from one place to another. Plainly the cause of this motion is the efficient cause of the body doing the pushing. So earthly things will remain at rest so long as they can't participate in some natural motion (they are prevented from falling, for example) and so long as no efficient external cause forces them into motion.

But Aristotle realizes he now has a problem. For a rock thrown by us continues in motion long after it is separated from the hand which, when it was in the process of being tossed, was the efficient cause of its forced motion. Aristotle speculates that somehow or other the motion of the hand is communicated to the air. This motion in the air continues to be propagated, and the communicated motion of the air remaining in contact with the projectile serves as an ongoing efficient cause keeping the projectile in motion. Such views were apparently not original with Aristotle, since he considers more than one theory involving the air as the means of effecting the continuing motion of the projectile, rejecting the less plausible of them and accepting, tentatively, the most plausible.

Aristotle is aware that the medium through which something is traveling can resist its motion (leaving, of course, the motive power of the air in the case of projectiles quite peculiar). In several places he asserts some proportions that are as close as he comes to a formal law of motion. Let the force exerted on an object be F, the resistance of the medium be R and the distance traveled in the motion in a time T be S. Then Aristotle's views can be summarized as the claim that S divided by T (the speed) is proportional to F divided by R, assuming, that is, that the force is sufficient to cause any motion at all. One consequence of this view, along with the view that resistance increases with density, is that in a vacuum the speed of a forced object would be infinite. Since that is impossible, Aristotle maintains, there cannot ever be a vacuum.

2.4 DYNAMICS IN MEDIEVAL ISLAM AND MEDIEVAL LATIN EUROPE

One of the first trenchant critiques of Aristotle on motion was that of John Philoponus in the sixth century AD. Philoponus believed that the speed with which an object traveled was proportionate to the force impressed upon it (believing, falsely, then, that heavier bodies fell faster than light ones did). But he was clear that even in a vacuum speeds would be finite. One could summarize his view by saying that the time of travel would be the sum of two times, one given by distance traveled divided by force and the other by an additional time due to resistance. Only that additional time would go to zero in the absence of any resistance.

More importantly, Philoponus rejected, on good empirical grounds, any explanation of the continued motion of the projectile that invoked the mechanism of the surrounding air. He argued, instead, that the hand throwing the projectile "impressed" on the projectile an "incorporeal kinetic force," and that this force continued to generate the motion of the projectile until used up by air resistance and the weight of the projectile.

The Islamic scholar Avicenna took up these ideas of Philoponus with subtle modifications. He speaks of various kinds of "inclination" (*mail*) which an object can possess. Natural *mail* is the inclination that weight gives an object and causes it to be set into motion by its own inner inclination in falling toward the Earth. But forced *mail* can be impressed on an object by an external cause, as the hand impresses it on the projectile. It isn't clear whether "inclination" for him really is anything different from "impressed force" in Philoponus, but it is clear that Avicenna believes that in the absence of a resisting medium the inclination impressed on a projectile will remain and the projectile will perpetually remain in motion.

Abu ‘l Barakat also discusses the problems of motion. Unlike Avicenna, he takes it that several different kinds of *mail* can exist in an object simultaneously, so that a thrown object near the Earth can move under the joint influence of its impressed forced *mail* and its natural *mail* due to its weight. Unlike Avicenna, he thinks of forced *mail* as dissipating with time even in the absence of a resisting medium.

These Islamic ideas were known to the great philosophers of the later Middle Ages in the period of the reawakening of learning in the Latin west of Europe. Most of the earlier Latin philosophers, however, rejected the tradition following from Philoponus and the exponents of the theory of *mail.* The fourteenth century experienced a vigorous scholarly debate concerning the problem of projectile motion. Some philosophers picked up on the notion of an "impressed force." William of Ockham thought that no explanation for continued motion was necessary. But most important for future science was the doctrine of Buridan. He speaks of "impetus," which is a property instilled in the object by the projecting force. It is impetus that continues the motion once the external driving force has been removed. Like Avicenna, he thinks of impetus as permanent, unless dissipated by resistance. He even gives a measure of the amount of impetus in a moving object, taking it to be, like the later notion of momentum, proportional to the velocity of the object and to the quantity of matter contained in it. A great deal of the final notion of inertia is contained here, but not all of it, of course. For example Buridan, in common with many other later thinkers, thinks of circular motion as sustained by impetus just as is motion in a straight line.

The impetus theory had a continuous history following its introduction by Buridan, until it finally was replaced by the full-fledged notion of inertia in the great scientific revolution in dynamics of the sixteenth and seventeenth centuries. Often the theory took backward steps, as in Oresme's reversion to the notion of impetus that by itself dissipated away with time. But the surviving idea of something communicated to the projectile that sustained it in motion persisted and was highly influential on Galileo's thoughts about motion and its causes.

SUGGESTED READING

An excellent survey of ancient Greek astronomy can be found in Chapters 9–14 of Pannekoek (1961). A thorough treatment of the subject is given by Evans (1998). A very lucid exposition of the Ptolemaic system at an elementary level is given by Taub (1993). Barbour (2001) discusses Aristotelian dynamics at length in Chapter 2. Chapter 3 gives a clear and thorough account of the mathematical models of Hellenistic astronomy. Chapter 4 outlines the development of impetus theory from the Islamic *mail* to later European impetus theorists. Clagett (1959), Part III, Chapter 8 is invaluable for understanding impetus theory. Clagett's introduction surveys the history from Aristotle through John Philoponus and the Islamists to the medieval Europeans. This is followed by useful textual abstracts.

CHAPTER 3

The astronomical revolution

3.1 COPERNICUS

3.1.1 The Copernican heliocentric system

The great breakthrough that ultimately led to modern astronomy and cosmology came with Copernicus' heliocentric system for describing the heavenly motions. But practically nothing one can say about Copernicus, either about his system or about its origins, is simple.

For one thing, the Copernican Universe is not the modern one, but the ancient cosmology in many of its most important respects. It is still a finite world bounded by a sphere of fixed stars. For another, Copernicus was not started on his path to his great discoveries by seeking a system that could use the Earth's motion to provide a more unified account of the apparent motions of the Sun and planets. Rather, he was dismayed by the appearance of the device of the equant point in Ptolemy's system and sought a means of eliminating it from an account of the heavens. Copernicus was even more devoted than the Greeks to the view that all the heavenly motions must be described in terms of pure circular motions in which everything moved uniformly about the circle with respect to the actual central point of the circle. Of course, hierarchies of circles, for example epicycles whose centers moved uniformly on deferent circles, were permissible.

Copernicus published his work only very late in his life. Indeed, there is a story to the effect that he first saw a printed copy on his deathbed. It is possible that the long delay in the book's appearance was at least in part due to the apprehension on Copernicus' part of the animosity such a revolutionary theory would give rise to, on the part, on the one hand, of those who felt a moving Earth violated religious precepts, and on the part, on the other hand, of established astronomers dedicated to a geocentric system. When the book was published it did, indeed, contain a preface, actually written by Copernicus' student Osiander, that characterized the new system as a mere mathematical device rather than as a true picture of how the cosmos was actually constructed.

But it is also probable that the long delay in publishing the work was in part due to Copernicus' dissatisfaction with his system. We will note several respects in which the system is deeply flawed. There are some respects in which it is as complex as, or sometimes even more complex than, was Ptolemy's system. Even worse, although Copernicus' system brilliantly solves some puzzles about otherwise unexplained correlations of phenomena in the Ptolemaic scheme, it introduces very peculiar cosmic correlations on its own part that seem mysterious at best.

Copernicus notes himself that some ancient Greeks had posited motion for the Earth, although he knows of this only through vague references in Plutarch. He affirms the right of an astronomer to use such an earthly motion, so long as it is in a circle and uniform about the central point, to save the observable phenomena. His main influence is Ptolemy's *Almagest*, and a very large portion of Copernicus' work can be considered an attempt to transcribe the results of the Ptolemaic system into a system that is heliocentric, in which the Earth has both rotary motion about its axis and orbital motion about the Sun (or, rather, about a point near the Sun), and which is devoid of equants and has all of the planets also moving uniformly in circles of which the primary circle is a heliocentric orbit. There remain deep historical puzzles about the influence of other astronomers upon Copernicus. The most important of these is whether or not he had any acquaintance with a number of works of Islamic astronomy. He never refers to such work, but both of the two devices he uses to get rid of the hated equant in his system had previously been discovered in Islamic astronomy.

Like Herakleides and Aristarchus, Copernicus attributes the daily rising and setting of the stars to a rotation of the Earth on its axis. The account of all other apparent motions begins then with this apparent daily motion due to the Earth's rotation. The apparent motion of the Sun through its yearly cycle is explained, again just as in Aristarchus, by the motion of the Earth in a circle about a point located near to the Sun. Unfortunately, Copernicus takes the Earth's orbit to be the heliocentric transcription of Hipparchus' solar orbit about the Earth, just as Ptolemy had swallowed Hipparchus whole. This puts the center of the Earth's orbit (even if it really were a circle) twice as far from the center of the Sun as it ought to be placed. Much mischief follows from this in Copernicus' system.

The great success of Copernicus' system is his use of the orbital motion of the Earth about the Sun to offer a far more unified account of the planetary motions, including the notorious retrograde motions of the planets in their trip across the heavens, than could be offered by Ptolemy. The major purpose for which the epicycles were introduced by Ptolemy is now served by the changes in the apparent place of the planet against the background of the fixed stars that are due to the Earth's own orbital motion. Although the idea of the Earth moving around the Sun had been speculated upon by the Greeks, and although the motion of the minor planets, Mercury and Venus, and perhaps that of the other planets as well, around the Sun had also appeared in ancient astronomy, it is Copernicus alone who realizes how a unified heliocentric system for Earth and planets alike can account for the aspects of planetary motion most notoriously difficult to explain.

There is even more. In Ptolemy the correlation of the motions of the major planets, Mars, Jupiter and Saturn, in their epicycles with the position of the Sun as observed from the Earth was quite mysterious. Why was the direction of a planet from the center of an epicycle the same as the direction of the Sun from the Earth? For the minor planets there was the equally mysterious fact that the center of the planet's epicycle was in the same direction from the Earth as was the Sun. Realize that in the case of the major planets the deferent circle represents the planet's own orbit around the Sun and the epicycle represents the Earth's motion around the Sun projected on to the planetary orbit, and that for the minor planets the deferent circle is the projection of the Earth's orbit and the epicycle the representation of the planet's orbit around the Sun, and these correlations of planetary and solar positions become trivial and obvious.

In the Ptolemaic system, each epicycle needed to be postulated independently. Now only one circle is independently posited for each planet. The other circle is fixed by the selfsame Earth orbit in each case. In the Ptolemaic system only the ratio of size of epicycle to size of deferent circle is determined. But now, given the fixed ratios from Ptolemy and the determinate value for the distance of the Earth from the Sun, the absolute value of each planetary orbit is fixed, as is the definite order in which the planets are arranged outward from the Sun. Originally, in the Copernican system this could only be considered an "aesthetic" virtue, since, of course, none of the actual planetary distances could be measured. Later, with the invention of the telescope, one can see the phases of the planet Venus, phases that are interpreted in terms of the planet's orbit around the Sun and the position of the Earth in a manner similar to the explanation of the phases of the Moon. This allows the use of trigonometry to determine the relative distance of Venus and the Earth from the Sun, and figures are obtained in conformity with those demanded by the Copernican system.

Is Copernicus able to rid his system of the hated equant? Yes, he is, but here one of the great mysteries of the history of science arises. For it was just the ability to dispose of the equant in favor of a system of epicycles that the Islamic astronomers had discovered. The two methods employed by Copernicus to abolish equants are just those discovered earlier. Quite possibly Copernicus had independently rediscovered these geometric devices.

If Copernicus' system was such a brilliant success, why did he himself have such grave doubts about its adequacy? We have already noted the errors in the Earth's orbit imported into Copernicus from Hipparchus by way of Ptolemy. Since the apparent positions of all the other planets depend upon the Earth's position, this imports descriptive difficulties into characterizing other planetary orbits as well. What is more, following

Ptolemy, Copernicus refers motions not to the actual position of the Sun, but to its *mean* position, the position it would have if taken to move uniformly around the earth. Each planet travels an orbit that is centered at the deferent point. If one referred all motions to the physical Sun, one would expect all the lines of apsides of each planet, that is the lines that connect the Sun to the deferent point for each planet, to intersect in the Sun. But Copernicus has them intersect in the mean Sun.

When he tries to get the orbits of the minor planets correct, the errors become egregious, and Copernicus' attempts to correct for them strange indeed. To get the longitude of Venus correct, he needs to have the center of its circular orbit move itself in a circle, a movement whose period is correlated with the time taken by the Earth to make a complete orbit of its own. It is worse yet with Mercury. And to get the latitudes of the planets correct he needs to have the planes of the planetary orbits tilt back and forth, again in a time that is determined by the phase of the Earth in its orbit. So having got rid of the mystery of why the planets have their motion "magically" correlated with the motion of the Sun about the Earth, Copernicus is stuck with the worse mystery of why the planets in their orbits about the Sun undergo motions that are even more magically correlated with the seemingly independent motion of the Earth in its orbit around the Sun! These difficulties are the result, once again, of referring motions to the mean Sun as in Ptolemy.

In addition to all of this there is the fact that numerous errors of observation used by Ptolemy remain uncorrected and taken as gospel by Copernicus. One thing is clear. Any naïve claim to the effect that Copernicus' system should have been seen by his contemporaries as obviously superior to the Ptolemaic system because it was "simpler," or "employed fewer epicycles," must be taken with many grains of salt. Copernicus had, indeed, made one of the greatest imaginable discoveries in the advance of descriptive astronomy. He had, indeed, "simplified" one vital aspect of Ptolemy's system, and he had done so in a manner that got things exactly right. It was, to be sure, the single motion of the Earth as observing platform in its orbit around the Sun that necessitated the need for an apparently independent dual circle system for each planet in Ptolemy's system. Copernicus was much less successful, however, in making any advance on Ptolemy in understanding the other reason why the motion of the planets is such a complicated matter. This is the fact that the planets do not, in fact, travel with uniform angular speed in circles about a center, whether that center be the Earth, the mean Sun, or even the physical Sun. It is not so much the deviation of their elliptical orbits from circular shape that causes the problems, but the fact that the speed of the planets in their orbits is not at all constant, being faster the closer to the Sun the planet is in its orbit. It remained for Brahe and Kepler to straighten all that out later.

3.1.2 Copernicus' defense of his system

Copernicus offers a short defense of his contention that the Earth is in both rotary and orbital motion against the most natural objections to those claims.

His cosmology, as mentioned, is still quite in the ancient format, with a stellar sphere surrounding a finite Universe. But now the Sun is taken as being at the center of that Universe. But why do we not see the obvious consequences of the Earth not being at the center and being in motion? Copernicus agrees that the motion of the Earth in its orbit causes us to have a different view of the planetary motions than we would were the Earth not orbiting. We see, for example, the retrograde motions whose cause is really in the Earth's own orbital motion. Why do we not see any effects on how the stars appear to us due to the fact that the Earth is, first of all, not centered in the Universe, and, second, in orbital motion about that center? If the Earth is not central, lines through the center of the Earth should not bisect the stellar sphere into equal halves, as they seem to do. And if the Earth is changing its position over the year from one place to another relative to that central point, there should be an observable change in angular position of individual stars as observed from the Earth, the phenomenon later called stellar parallax. The answer must be that the sphere of the fixed stars is vastly more distant from the central object, the Sun, than is the Earth or any of the planets. It is only this vast distance that makes any observable effects due to our Earth's off-center and moving position with regard to the center of the stellar sphere undetectable by us. That answer is, as a matter of fact, correct, but one can well imagine Aristotelian contemporaries of Copernicus finding it ad hoc.

The Aristotelians took the Earth's stationary, central position to be associated with its power to attract matter, or, rather, with its nature as the location of the central, natural place of Earthly matter. But, suggests Copernicus, perhaps each heavenly body is the natural place of its own substance, so that there is no inference to be drawn from the tendency of the Earth to gather earthly matter regarding its centrality or immobility. The Aristotelians may object to any theory in which the Earth suffers more than one motion, as the Earth does in both rotating and traveling in its orbit. But any account of planetary motion that restricts itself to uniform motion in circles will require multiple circular motions for the planets, so compound motions are endemic to cosmology.

The hardest problems encountered by Copernicus, though, deal with the rotation of the Earth. If it fully rotates in one day and has the substantial size it is known to have, its surface must be moving at quite high speeds away from the poles of the axis of rotation. Why doesn't it fling things off its surface? Why doesn't it tear itself apart? Why aren't the clouds left behind in its rotation? Why, if we toss an object into the air, does it come down where we tossed it?

First of all, Copernicus argues, the problems involved in the Earth's rotation pale in comparison with the problems faced by someone who has the stellar sphere in rotation. The planets move more slowly the further they are from the center of the Universe. Why should the stellar sphere move so quickly when so far away? And why wouldn't the sphere tear itself apart? Wouldn't it be forced to expand without limit if it was rotating so quickly, leading to the absurdity of ultimately infinite speeds for the stars? Copernicus also counters Aristotelian claims about there being nothing outside the stellar sphere by arguing that this would make the expansion of the sphere all the easier.

More important arguments, though, try to explain why we don't experience the predicted effects of the rapid rotation of the Earth, rather than simply trying to argue that those who believe in a rotating stellar sphere are afflicted with worse problems. Perhaps all of those things that belong to the Earth, including clouds and objects tossed in the air, share in the "natural" motion which is the Earth's rotation. As they continue, even when unattached to the Earth, to share in its rotation, none of the predicted effects due to being "left behind" will be experienced. An object tossed in the air begins to fall. But as it falls it has a dual motion. It has a straight-line motion of fall, because it was displaced by the tossing; but it continues to have the natural circular motion it shares with all things belonging to the Earth. As a consequence, it comes down where it was tossed. Why doesn't the Earth tear itself apart? Well, its rotary motion is a natural motion, and natural motions are directly "contrary to violence." Such predicted disintegration, being violent, would not be expected to be the consequence of a "natural" rotation such as that of the Earth. So, unlike the potter's wheel which shoots off the clay upon it if spun rapidly, the rotating Earth being in natural rotation cannot be expected to fling off its bits and pieces as it spins.

These suggestions for evading the arguments of those who find dynamical reasons for rejecting the rotation of the Earth are certainly not correct. Nor, as we shall see, are those of Copernicus' greatest defender, Galileo. But they are errors that lead to the truth, which the Aristotelian errors do not.

3.2 BRAHE AND KEPLER

3.2.1 Brahe's observations and Brahe's cosmology

Tycho Brahe's lifetime of careful observations of the heavenly motions finally provided the kind of accurate data, purged of errors accumulated before and after Ptolemy, that was needed for the construction of an accurate mathematical model of the heavens. He also confirmed by careful observation that comets were heavenly, not atmospheric, phenomena, and that the orbit of a comet pierced what would have had to be the crystalline spheres of the ancient physical models of the heavens, putting that old hypothesis finally to rest.

Brahe also had an ingenious model of the cosmos. He proposed it, but never worked it through mathematically in detail. Like all the older models, it assumed a sphere of fixed stars. Like in the ancient models, the spherical Earth was at the center of the cosmos and at rest, avoiding the problems of the absence of parallax for the stars and of the absence of dynamic consequences of the Earth's motion faced by Copernicus. But, possibly like some ancient models that may have existed, he had, as did Copernicus, all of the planets orbiting around the Sun. The Moon circled the Earth as did the Sun, but all the other planets traveled about the Sun.

The model never "caught on" at the time, and today we think of it as a quaint transient proposal in the history of science. But it would be wise to consider how, kinematically, the system can be made to give exactly the same results as Copernicus' system, and, given the then existing astronomical data (no observed parallax for the stars) and the then known dynamics, how neatly Brahe's system gained the advantages of Copernicus (taking the main epicycles to be replaced by the Sun's motion about the Earth as Copernicus had eliminated them in terms of the Earth's motion about the Sun), without paying its penalties. Of course, without a mathematical development, nothing in Brahe's model by itself helped in ridding astronomy of all the peculiarities and errors that infected both the Ptolemaic and the Copernican systems.

3.2.2 Kepler's laws of planetary motion

Brahe's great successor, Johannes Kepler, combined a deep desire to offer a physical account of the motions in the heavens with an extraordinary power of insight, combined with the seemingly infinite energy and dedication needed to lift the ability of descriptive mathematical astronomy to capture the observations to a new height. Kepler's work often has "mystical" elements to it that make him seem very pre-modern indeed. It is also true that his overall cosmology still has the distant sphere of fixed stars, making him seem quite old-fashioned. On the other hand, the way in which Kepler combines thoughtful and imaginative mathematical model building with trenchant physical speculation makes him seem to us much more a modern scientist than does Copernicus. Kepler was a convinced Copernican, arguing that Copernicus' successful accounting for all of the major epicycles and their correlations to the position of the Sun in terms of the Earth's motion was a gain never to be lost. And if the Earth could orbit the Sun, why should it not rotate as well?

One early insight of Kepler's underpinned all of his future results. Taking the Sun as the physical source of the planetary motions, he referred all motions to the position of the *true*, physical, Sun, rather than to the mean solar position used by Ptolemy and Copernicus. In a fundamental application of this idea, he took each planet to be moving in an orbit confined to a fixed plane, with all the planes intersecting in the Sun. This allowed him, after much hard work in finding for Mars where that plane intersected the plane of the Earth's orbit and at what angle, to solve the vexing problem of planetary latitudes in a simple way, ridding the Copernican system of all the weird planetary motions in latitude correlated to the Earth's motion.

Kepler then set himself to work on the longitude problem. The first thing he did was to get an improved orbit for the Earth about the Sun, thereby ridding astronomy of the perplexities introduced by Hipparchian errors carried onward by Ptolemy and Copernicus. Basically, Kepler introduced the same equant device into the Earth's orbit that Ptolemy had used so effectively for the planets. Next Kepler realized that a planet's velocity in its orbit increased with closeness to the Sun. At first he tried to represent this by a law that had the velocity of the planet inversely proportional to its distance. He found this hard to work with for making predictions, and approximated it by a law that was more tractable, the law that later became known as Kepler's Second Law. This was the law to the effect that the area swept out by a line drawn from planet to Sun was the same in a given fixed time interval no matter where the planet was in its orbit. Only later did Kepler realize that this was the correct law. (It agrees with the inverse distance law only when the planet is at its nearest to or farthest from the Sun.)

Kepler then worked assiduously to get the parameters for the orbit of Mars correct in longitude. Even with the new speed law and with a circular orbit whose center was displaced from the Sun, he couldn't get exactly the right results to the degree of accuracy needed to fit Brahe's data. He needed to "squeeze in" the orbit when the planet was at right angles from the line of the apsides, namely the line containing the perihelion and aphelion points, the points on the orbit when the planet was at its closest to and farthest from the sun. He tried oval shapes, and, once again, thought of approximating these by the mathematically tractable shape of the ellipse, about which much had been known beginning with Appolonius' work on the conic sections. To his great wonder and joy he found that he could eliminate the last errors in the orbit by taking the orbit to be an ellipse with the Sun located at one of its foci. The statement that the planetary orbits are ellipses with the Sun located at a focus became known as Kepler's First Law.

Kepler's physical explanations of the planetary motions mostly relied upon the idea that the Sun exerted a force on the planets that "drove" them about in their orbits. Here analogies with the force of magnetism, which had recently been described by Gilbert, pervaded. One idea was that the Sun rotated, and sent out force from itself that dragged the planets along with the Sun's own rotation. Kepler took the fact that planetary motions were slower the further from the Sun the planet was as an indication that the force exerted by the Sun dropped off with distance. He tried several laws for this diminution. In one he wonders whether the planet's velocity decreases inversely with distance from the Sun. In another he takes this to be only the transverse velocity of the planet instead. None of this works.

What he does discover, however, is Kepler's Third Law. This states that the time taken to make a complete orbit for a planet increases as the three-halves power of the inverse of the distance. So if we have two planets with their orbit times, t_1 and t_2, and their mean distances from the Sun, r_1 and r_2, the law states that $(t_2/t_1)^2 = (r_2/r_1)^3$. Later this law provided the single most important clue regarding just how the force from the Sun that does affect the planets really does diminish with distance.

SUGGESTED READING

Chapters 18–24 of Pannekoek (1961) provide a clear survey of the history of the astronomical revolution. Chapters 5 and 6 of Barbour (2001) will introduce the reader in a very clear and direct way to the more technical aspects of Copernicus and Kepler. Kuhn (1957) provides a survey of the place of Copernicus in the revolution in astronomy. Swerdlow and Neugebauer (1984) is a profound and thorough study of the details of the Copernican system, warts and all.

CHAPTER 4

Precursors to Newtonian dynamics

Many threads were finally woven together in the great Newtonian synthesis from which all further developments in classical dynamics followed. In this chapter we will outline a few of the major contributions to dynamics that followed the Copernican revolution in astronomy, but preceded the final accomplishment of a full theory that we can recognize as classical dynamics. Three names dominate this early work on dynamics, those of Galileo, Descartes and Huyghens, but, as we shall see, important contributions were made by less well-known figures as well.

4.1 GALILEO

Almost everyone would give Galileo credit for initiating the great revolution in dynamical theory that ultimately led to Newtonian classical dynamics. But, as is the case in the work of many originators of a new science, there is no simple way of characterizing Galileo's contribution. His work, while being innovative in a revolutionary way, retains much that in retrospect seems quite conservative in its nature, borrowing much in the way of concepts, views and arguments from his predecessors. His exposition is sometimes quite informal, presented in the form of a charming dialogue in some cases. This sometimes makes it rather difficult to say exactly what Galileo believed, since, in some crucial cases, he seems to hold a number of distinct opinions simultaneously. In some crucial cases we would now be inclined to think that Galileo got things very wrong indeed. But even in his errors he provides deep enlightenment.

Galileo was schooled in impetus theory, and in his early years held to the unfortunate version of that theory that had impetus dissipating away even in the absence of friction or other external resistance to motion. He also held, as did many others, that falling bodies developed speeds of fall in proportion to their weight. Later he rejected both of those errors.

Galileo was an avid experimenter, and developed quite ingenious methods of bringing nature under sufficient control that even with the crude devices at hand (such as timing processes by crude clocks or even by rhythmic tapping) quite accurate experimental determinations could be carried out. Galileo was the first to use the telescope to make important astronomical discoveries, such as sunspots and the moons of Jupiter, and his work did much to break down the old Aristotelian distinction between the unchanging nature of the heavens and the changing nature of sub-lunar phenomena.

Galileo was a fervent advocate of applying mathematics to the description of nature, and brought the use of geometric picturing and reasoning out of the heavens, where the astronomers had applied it, into the realm of earthly motions. In characterizing motions he is usually content to give a mathematical description of the motion, and to avoid extensive discussions of "causes" that may lie behind the motions. Where explanations are given, they are often of a quite Aristotelian sort, invoking natural places and natural motions.

Galileo was also able to "think away" disturbing features of a situation, going to an ideal case, even when those disturbing features were not really experimentally removable. In his theory of falling bodies, for example, he is willing to infer how things would behave if the irremovable resistance of the air were absent, and in his discussions of moving objects he is often asking the reader to join him in thinking away the disturbance of friction in order to discern how things would behave in its absence.

Galileo embarked on a systematic study of the nature of the motion of bodies in free fall. This problem had been touched upon by many of the earlier students of dynamics, but with few conclusive results. Galileo made two discoveries, both of which were fundamental for future science.

First, by dropping objects of different sizes and composed of different materials from a tower, he noted that, differences due to differential air resistance being neglected, the objects reached the ground at the same time. Thus "gravity," whatever it was that caused unsupported objects to move toward the Earth's center, acted in such a way as to induce the same motions in all objects, irrespective of their size or composition. Later Newton was to recognize the fundamental importance of this result. He tested it to great degrees of accuracy using pendulums made of different materials, and it plays a crucial role in his formulation of the law of gravitational force. Later still, Einstein was once again to recognize the importance of Galileo's great discovery. In his hands it became one of the guiding intuitions behind the discovery of the general theory of relativity.

Galileo also discovered the fundamental quantitative law of falling bodies, that freely falling bodies have their velocities of fall increase by equal amounts in equal times. Galileo cleverly "slowed down" the falling process by rolling balls down inclined planes, relying on a number of insights to connect the motion down the plane with the motion an object would have when falling completely unconstrained. His basic experimental result is that the distances covered in free fall increased with the square of the time of the fall. It was from this result that he obtained, after several years of thought, the result that the speed of the falling object was itself directly proportional to the time of fall.

This result later became the fundamental particular case from which all later generalizations connecting force and motion are derived. When Huyghens derives the formula for the amount of force a string must exert on an object to keep it moving at constant angular speed in a circle, it is from Galileo's laws of falling bodies that he obtains his result. And it is Galileo's result, broadly generalized, that later becomes Newton's Second Law of Motion in his *Principia*.

Galileo's work on the persistence of motion, the work that led eventually to the final version of inertial motion, was inspired by his fervent Copernicanism. How could the Copernican doctrines be defended against the attacks against it which were based on the claim that a rapid rotation of the Earth would have obvious dynamical consequences on the Earth that we do not observe?

Much of Galileo's reply seems to be a "filling in" of the brief remarks made by Copernicus himself. There are natural motions in the world. One of these is the rotation of the Earth. All earthly things share in that rotation and, hence, we see no consequences of relative motions of things (clouds, birds, tossed objects) relative to the Earth. A falling object also has a natural straight-line motion. Its true motion is a compound of the circular motion it shares with all earthly things and its own direct fall. Hence, an object dropped from a tower will fall at the foot of that tower. So there is one kind of "persistent motion" which is the natural circular motion shared in by all earthly things.

But Galileo also observes that, if we roll a ball down an inclined plane, it will rise up another sloped plane to the height at which it was first released. If the second inclined plane has a very small slope, the ball will travel a large distance before reaching its original height (always ignoring friction, of course). Ought we not to assume, then, that, if the second plane is horizontal, the ball will roll on forever? So there is another sort of persistent motion, which is motion in a given direction along the horizontal on the Earth's surface. Such motion will persist without ceasing so long as it is not impeded by resistance or friction. It is noteworthy that, when Galileo speaks of these persistent motions, he speaks of them as not needing any cause to keep them in operation, rather than as being caused by some "impetus" that they acquired earlier.

Galileo also notes the persistent motion of a rock flung from a sling, or a projectile fired from a cannon. Here the motion persists in a straight line without losing any of its speed, again unless interfered with by, say, a resisting medium. Allowing for motions compounded of two simple motions, Galileo considers the motion of a projectile fired near the surface of the Earth. Its motion being a compound of its persistent straight-line, constant-speed motion and its accelerated motion toward the Earth as it falls, he calculates that its overall path will be parabolic.

In defending Copernicus Galileo also makes major breakthroughs in discussing the relativity of motion. On a ship moving smoothly, he says, a ball dropped from the mast will land at the foot of the mast. (He notes that he hasn't actually done that experiment, but is sure of what the result would be if he did!) A horseman can toss a ball up in the air and repeatedly catch it. It isn't left behind as he rides. The dropped and tossed objects share in the motion of ship and rider, and so are not left behind, any more than are clouds, birds and tossed objects by the Earth. Galileo's most dramatic presentation of "relativity" is in his discussion of what would happen if one performed any dynamical experiment whatever when confined in the hold of a smoothly moving ship. Nothing one did could tell one that the ship was engaged in any one uniform motion or another, rather than being at rest at the dock.

In the claims about the straight-line uniform motion of projectiles and in his discussions of the absence of dynamical effects of uniform straight-line motion of the laboratory, Galileo is onto core fundamentals of classical dynamics, and of other theories (special relativity, for example) as well. But it is historically important to notice that he also has the idea of natural circular earthly motions and a kind of "circular inertia" as well. It took some time for early science to disabuse itself of that idea. Other Galilean arguments are also wrong-headed. There is, for example, a quite confused argument to explain why objects are not thrown off by the rapidly spinning Earth.

Another wrong-headed argument is used by Galileo as what he takes to be one of the most conclusive arguments for the Earth's rotation. He believes that that rotation can be dynamically exhibited, even if most of the dynamic effects predicted by anti-Copernicans do not occur. He has a theory of the tides that relies upon the fact that on opposite sides of the Earth the velocity of the Earth's surface due to its rotation around its axis and its velocity due to its orbital motion are, on one side, added to one another, and on the other side act in opposite directions. Drawing an analogy from the way water sloshed in barges carrying fresh water to Venice when the barge was suddenly accelerated, he thought he could explain the tides as a kind of sloshing due to the Earth's compounded motions. Galileo, therefore, thought of the Earth's motions as real, and not as a mere astronomer's device to save the phenomena, or, indeed, as some merely relative kind of motion.

4.2 DESCARTES

The contrast between Galileo and René Descartes could not be starker. Galileo rarely offers physical explanations for how things behave, but directs himself to offering detailed mathematical descriptions of motions as they are experienced in nature. Descartes offers grand physical, cosmological and metaphysical accounts of everything in the Universe. Among the things he accounts for is motion in all of its guises. But it is rare for Descartes to give us any detailed mathematical description of some important dynamical process. And when he does do so, he often goes wrong in extraordinary ways! This is even more surprising when we note that Descartes was sufficiently a mathematician of genius that we owe to him the origins of analytic geometry.

Descartes' philosophical methodology is rationalism. Just as the truths of Euclidean geometry are derived by pure logic from (allegedly) self-evident first principles, the geometric axioms, so all the other truths about the constitution and action of the Universe should be rationally deduced from "clear and distinct" ideas.

Descartes' grand metaphysics posits a world made of two substances: mind and matter. Each substance is characterized by a single essential property whose nature and variation characterizes all that is and all that changes in the world. For mind the property is "thought" (consciousness?). For matter the essential property is spatial extent.

Whereas Aristotle contemplated many changes in matter, for Descartes there is only one change, motion or change of place. Whereas Aristotelians would give the causes of changes by reference to special properties of matter, properties that might even be "occult" in the sense of being hidden from our direct observation, for Descartes all change, that is, all motion, is to be explained by invoking previous motion. To explain why bodies fall to the center of the Earth, or to explain why the planets move as they do, we must refer to the spatial properties of all the relevant matter contiguous to the object in question, in particular to all of the motions of that surrounding matter. For example, consider the problem of the motion of the heavenly bodies. This Descartes explains by positing a cosmos filled with a subtle matter, the plenum. The plenum is continually in motion, and it moves the planets immersed in it by its motion, just as a cork moves with the water in which it floats. The kind of motion that explains the behavior of the planets is the motion of a fluid vortex or whirlpool with the Sun at its center. It is this perpetual swirl that drives the planets in their courses. But in Descartes all we get is a picture (indeed, only a two-dimensional one, not the three-dimensional one that would be appropriate) of how all of this works. No effort whatever is made to show how such a physical model could possibly get the motions of the planets as described by Kepler right. Indeed, as Newton later showed in detail, it cannot.

We ought to notice at this point, for future reference, that Descartes gives us here a prime example of one of our major methodological themes. Descartes' account of the causes of motion is in terms of each motion being generated by contact with prior contiguous motion. This does not just amount to a physical causal theory, but, for Descartes and his followers, constitutes a general principle of what the structure of an explanation of any kind must be like. To advert to hidden explanatory properties, to occult qualities, is not just to get the explanation wrong, it is to indulge in offering *pseudo*-explanations that are not really even of the proper form to be contenders as scientific explanations. This is a matter to which we will return in some detail in Chapter 6 when we consider the philosophical issues raised by the Newtonian synthesis.

We should note, furthermore, that Descartes' cosmology is also a final complete break with the finite cosmos of Aristotle. Descartes' Universe is an infinitely extended Euclidean three space filled with the swirling plenum. And, like Giordano Bruno, he thinks of the stars as all suns, each likely to have its own world of planets swirling in its vortex.

Descartes' theory of the cause of motion has another aspect that is of extraordinary importance. Here Descartes fixes on something that is in essence correct and of vital importance in the future of dynamics. But in matters of mathematical detail almost everything goes wrong in the Cartesian account. For Descartes matter is conserved. All that there was at the beginning of the Universe is still present, and no new matter has been added. But motion is also conserved. Each change in motion is the result of a transfer of motion from one piece of matter to another. This principle, together with the principle of the persistence of motion when not interfered with that we shall discuss shortly, is at the core of Descartes' theory of how motions change.

The changes of motion that suggest this principle most clearly are those we call collisions. One ball in motion collides with a ball at rest. The second ball then moves, and the motion of the first ball changes in the process, sometimes coming to a complete halt. The fundamental principle espoused by Descartes is that whatever motion one object has gained in such an interaction is motion that is lost by the other object. The sum totality of motion in the Universe remains always conserved.

The problem is that Descartes doesn't get the right idea of exactly what is conserved in such a process. We noted earlier that Buriden already had the notion of a quantity of motion as the product of amount of matter and speed of motion. Descartes would like to use as a measure of quantity of matter the volume of the object, since he takes only spatial extent as being the true essential property of matter. It won't work. Some notion of what later came to be called "mass" as a primitive measure of "quantity of matter" is essential. Worse yet, Descartes takes the measure of motion to be the speed of the object rather than its velocity, where the latter is used to indicate a simultaneous determination of speed and the direction of that speed, i.e. a vector rather than a scalar quantity. Even in the simplest one-dimensional case of collisions you need to take account of the direction of motion of the colliding objects in order to get the conservation of what later came to be called momentum to work out right.

Descartes also proposes a number of detailed laws of collision. Suffice it to say that they are a mess. They seem to have been derived from the work of Descartes' early companion with whom he developed many of his ideas,

Isaac Beeckman. But whereas Beeckman thought of some of the laws as applying to what are now called inelastic collisions, Descartes thinks they apply to elastic collisions. Several of the laws contradict each other. Others give rise to results that are manifestly false, as can be seen by performing the simplest laboratory experiments. To make matters worse, Descartes assures those skeptical of the truth of the laws that they follow from clear and distinct ideas, and that any appearance of falsification in experiments is due to the failure of the experiments to meet the necessary ideal conditions such as lack of all friction.

Descartes' most important permanent contribution to dynamics is that he gets what came to be called the law of inertia exactly right. An object not interfered with by anything external to it will persist in motion with constant speed in a straight line. Getting to this point was, as we have seen, a complicated story. It is actually even more complicated than has as yet been related. The path led from Aristotle, through John Philoponus, through the Islamic theorists of *mail*, through Buridan and the impetus theory and then up to Galileo. We went from external force needed to sustain motion, to the idea that something was impressed upon the projectile that sustained its motion without the need for anything external to do the job, to the idea in Galileo of motion that naturally persisted on its own.

Buridan and Galileo, though, clearly believed in self-sustaining circular motion as well as motion in a straight line. Beeckman also believed in circular as well as linear inertia. Curiously, prior to Galileo J. B. Benedetti, an impetus theorist, had already made the assertion that only motion in a straight line was so self-sustaining. He used the same example that so impressed Descartes, a rock flung from a sling. Benedetti then went on, however, to claim that, therefore, such things as the perpetual spinning of a disk which suffered no resistance were impossible – which is not true.

The biggest hurdle to getting the truths of dynamics was the need to realize that not motion, but change of motion, is what required the imposition of interference from the outside world. This the impetus theorists Galileo and Descartes accomplished. Understanding that perpetual rotation of a rigid body was possible, but that this did not require that such motion be unforced as is true inertial motion, was the second difficult hurdle to overcome. When a disk spins, the point masses in it do suffer forces, the internal forces that hold the disk together and keep the points from flying off in straight lines. But these forces do no work, serving only as constraints, and play no role in changing the perpetual rotary motion of the disk.

But, in any case, Descartes got it right. Reflecting on a rock in a sling imagined detached from the sling at some point in its motion, Descartes asks what else ought we to expect but that the rock will continue in its instantaneous motion at the time of detachment, that is, moving off in the straight line at a tangent to its previous circular motion with a constant speed of the value it instantaneously had at the moment when it was detached.

Finally there is Descartes' doctrine of the very nature of motion. What does it mean for an object to be at rest or to be in motion, or to have some specific kind or degree of motion? For Aristotle there was the standard of rest in the fixed, immobile Earth at the center of the Universe. But for Descartes there is no fixed Earth and no sphere of fixed stars. There is only infinite space, which is the same everywhere. Of course, the space is filled with the plenum. But that is everywhere in different kinds of vortical motion, and cannot supply a standard of rest either.

Descartes solution is to deny that any single standard of rest is required. What is it for something to be in motion? It is merely that the thing concerned is changing its place relative to something else. An object can be at rest relative to the Earth, but, at the same time, in motion relative to the Sun. Is it really at rest then or really in motion? There is no point to asking that question. Motion is purely relational, x is moved relative to y. Furthermore, if x is moved relative to y, then y is moved relative to x! Of course there are cases where it is natural to fix on one object as the standard of rest for some purpose, in which case, with that standard presupposed, we can simply ask whether some object is moving or not. But such a choice is one of practicality or convenience.

In one of Descartes' most curious arguments, Descartes, the fervent Copernican, finds himself able to avoid the dangers of conflict with the church that made Galileo recant his doctrine of the Earth's motion. Our ordinary standard for saying whether something is moving or not is often whether it is in motion relative to its immediate surroundings. But the Earth is not moving relative to the plenum material immediately surrounding it, which drags it in its vortex around the Sun. So we ought to say that the Earth is at rest!

Descartes has, in any case, clearly propounded the relationist theory of place and of motion. Things have location only relative to other things, and undergo change of location, that is motion, only relative to other things chosen as standards of rest. There is no such thing as "being at rest" *simpliciter*, or being in motion *simpliciter*. That this doctrine, accepted as we shall see by Huyghens and brilliantly expounded by Leibniz, is incompatible as it stands with the principle of the persistence of straight-line motion at constant speed, that is with the law of inertia, remains something for Newton to make abundantly clear.

4.3 HUYGHENS

Descartes presents grand physical schemes and few detailed mathematical descriptions of motion. Christiaan Huyghens also has grand schemes in mind, in fact Cartesian schemes of which he became enamored early in his life. But his own attention is primarily focussed on the problem of the exact mathematical descriptions appropriate to crucial dynamical problem situations. In two cases, that of collision and that of the force needed to keep an object moving in a circle, he is eminently successful in his quest. What is even more striking to the contemporary reader is the degree to which Huyghens invokes methods of thinking that remain as essential to theoretical physics today as they were for Huyghens. His work has, throughout, a "modern" feel to it that is quite remarkable.

We noted that Descartes had taken on the problem of collisions and made a hash of it. Of his seven laws, only the first is correct. But Huyghens takes off from that one true law to derive, by means of posited fundamental principles, an extended class of correct mathematical descriptions of collisions.

Descartes' first law held that, if two bodies of equal size approached each other with equal speeds, they would rebound from each other with the same equal speeds, which is certainly true for elastic collisions. What is the "size" of an object? We saw that Descartes thought of it as volume, which won't do. Huyghens takes size to be measured by weight. Since what is later called mass by Newtonian physics and weight are proportional in a fixed gravitational situation, no error is induced by Huyghen's decision.

Huyghens then takes Galileo's observation that physics all appears the same in any uniformly moving frame of reference and makes astonishing use of it that goes far beyond anything Galileo had contemplated. In doing so he provides us with the first example of that recurrent device used in mathematical physics of taking a process as described in one reference frame and transforming its description to another frame moving with respect to the first.

Let a collision of objects of equal weight moving toward one another with equal speed take place as observed in one uniformly moving reference frame. Describe the result of that collision by Descartes' first, correct, law. But now describe what happened in a frame moving uniformly with respect to the first. The speeds of the colliding particles will no longer be identical before or after collision. But they can easily be derived by just adding the speed of the relative motion of the reference frames to the speed of the particles. By this means a one-parameter infinite class of collisions can be correctly described. The specific result here is nowhere near as important as the principle of relativity invoked and the method of transforming descriptions between frames that is introduced.

A little later Huyghens takes up the problem of the elastic collision of two bodies that are unequal in size. Huyghens takes from Descartes the principle of the conservation of motion, where the quantity of motion is given by the product of size (now weight) and speed. But Huyghens, correctly, realizes that it is directed speed, that is velocity, that is crucial, rather than the absolute value of the speed.

But he discovers another fundamental principle as well, indeed the first statement of a fragment of the general principle of the conservation of energy. Galileo had observed that an object that falls from a height and gains a certain speed has just enough motion in it to reach that height again (friction, etc., of course, being ignored). But the height from which an object falls is proportional to the square of the speed it finally obtains at the end of the fall. Huyghens then generalizes Galileo's result by positing that in the collision of a number of particles the sum total of height-obtaining ability is conserved. That is, that the sum of the products of mass times the square of speed for each particle over all the particles is the same prior and subsequent to the collision. This is, of course, the conservation of kinetic energy in elastic collisions. The two principles of conservation of momentum and of energy are, amazingly, enough to solve completely the problem of two-body elastic collisions, without any need whatever to worry about any details of the collision process.

Huyghens also notes an "admirable law of nature" to the effect that the center of gravity of a system of particles continues in a motion that is uniform and in a straight line throughout a collision.

Huyghens' other main contribution is to the problem of centripetal (or, confusingly, centrifugal) force. Suppose an object is constrained to move in a circle by means of a string attached to the center of the circle on the one end and to the object on the other. With what force does the string pull on the object? Indeed, what, exactly is meant by "force" here? Huyghens' answers to these questions are crucial to the future of dynamics for two reasons. First, his solution to the first question provides the key to unlock the problem of the dynamics of the motion of the heavenly bodies. Second, his answer to the second question provides the route from Galileo's work on the constant acceleration of falling bodies to the full Second Law of Newtonian dynamics.

Let us see how Huyghens deals with the second question first. Imagine, he says, an object above the surface of the Earth held in place by a string that prevents it from falling. What pull must the string exert on the object? Surely that should be measured by what gravity would do to the object were the string cut. But what gravity would do, if allowed, is to accelerate the weight. And it doesn't matter how the object would be accelerated after the string is cut (it might, for example, be guided down a curved ramp that constantly reduces its downward acceleration as it falls), it only matters what acceleration the object would feel at the very moment the string is cut. The "force" the string is exerting in preventing the fall, then, must be proportional to the instantaneous acceleration the weight would experience were the string cut. There is, of course, a constant of proportionality also that depends on the size (later we shall say the mass) of the object as well.

Now look at the object tied to the center of the circle by a string and moving uniformly about the circle. Imagine the string cut. The object will then move off on the tangent to the circle at the place the object occupied when the string was cut, and with the speed that it had in the direction of that tangent at the moment of cutting. That follows from the correct Cartesian principle of inertia. But then view the path of the detached object from the point of view of an observer at rest on the end of the string which is imagined to continue its circular motion at the rate it had the moment of the cutting. That observer sees the object fly away from himself along a curve. Huyghens shows by purely kinematic reasoning, that is, by considerations of geometry and time alone, that the object will be seen to move along a curve tangent to the radial string at the point of cutting. So, at the first instant, it is seen to move directly off in a radial direction. And he is able to show that at that first moment the distance the particle moves away from the cutting point increases as the square of the time, that is, it exhibits uniform acceleration in the radial direction. With a little geometry Huyghens can show that the acceleration has a magnitude that is either proportional to the square of the velocity of the particle and inversely proportional to its radial distance from the center of the circle, or is proportional to the square of the angular velocity of the particle and to the radius. That is, he derives the familiar formulae for centrifugal force, $F = mv^2/r$ or $F = m\omega^2 r$.

The exact structure of the formula for the magnitude of the force plays, as noted, the essential role in getting the law of gravitational attraction right in the later theory of the motion of the Moon and planets. The idea that force is to be measured as proportional to acceleration takes Galileo's fundamental discovery about free fall one stage further to the final lawlike association of force with change of motion or acceleration summed up in Newton's Second Law.

Huyghens also has some important things to say about the nature of motion, in particular about whether we can hold to Descartes' claim that all motion is merely relative. But it is best to reserve that until the discussion of Newton's fundamental attack on Cartesian relativism later on. Huyghens also made a major contribution to optics. One of the earliest exponents of the wave theory of light, he proposed as a fundamental principle of the propagation of light that a wave front could be constructed by taking it as the envelope to the surfaces generated by looking at each point on an earlier wave front as itself the origin of a new outward-spreading wave propagation. Curiously, this principle, called Huyghen's Principle, was later brilliantly applied by William Hamilton to a new fundamental formulation of dynamics. We will take up this optical work of Huyghens when we discuss Hamilton's use of it in mechanics later in the book.

4.4 OTHER PRECURSORS OF NEWTON

When we discuss Newton's philosophical views about space and time, we will note some influences on his work of earlier philosophers. Here, however, we are concerned strictly with a few additional antecedents of his scientific results.

The core influences on Newton's laws of dynamics have already been surveyed in our outline of the work of Galileo, Descartes and Huyghens and in summarizing the work that preceded theirs. Newton attributes to J. Wallis and C. Wren, who contributed alongside Huyghens to the theory of impacts, additional influence on his work, especially on the inspiration behind what became Newton's Third Law of Motion. Also some important contributions to the theory of the motion of the heavenly bodies were highly influential in the development of Newton's final theory.

The most important of these contributions was the work of J. Borelli, who, using the familiar image of the rock flung by the sling, argued that the motion of the planets could be explained by the planet's tendency to continue its motion at constant speed in a straight line and a force exerted by the Sun and directed to the Sun that pulled the planet in the Sun's direction. It seems, though, that Borelli also believed, with Kepler, in a tangential force exerted by the Sun as well that drove the planet in its forward motion.

As early as 1645 Boulliau had already suggested that the force exerted on a planet by the Sun diminished in proportion to the square of the distance of the planet from the Sun. But he took the force to be the unneeded tangential force. In works well known to Newton, Robert Hooke combined the correct parts of Borelli's analysis of the dynamics, tangential inertia and centripetal attraction, and took them alone to generate the orbital motions, illustrating his idea by means of experiments with pendulums. Finally someone had fully realized that no tangential driving force was necessary to the account. Nor was a force needed to keep the planet from falling into the Sun. Inertia and attraction were together sufficient to solve the problem. He also took up Boulliau's correct inverse square law, once again probably guessing at the relationship from the optical analogy. He posited that each cosmic body had such an attractive power that kept its satellites in their orbits, but never associated that force with the terrestrial effects of ordinary gravity or posited a universal gravitation as Newton did later.

Also very important was Edmund Halley's clear demonstration that, assuming this analysis of the dynamics, the law governing the force exerted by the Sun on a planet was indeed an inverse square law. Or, rather, he showed this to be valid if we assumed, as he knew was false from Kepler's work, that the orbits were simple circles about the Sun at their center. Here Halley used Huyghens' formula for the dependence of centrifugal force on angular velocity and radius combined with Kepler's Third Law of planetary motion that gave the correlation between the angular velocity of a planet and its mean distance from the Sun. For the first time the inverse square law was obtained not by guessing and analogy, but by a derivation (at least for a simplified case) from an established dynamical law and a well-known empirical lawlike correlation.

SUGGESTED READING

The texts by Galileo (1914, 1967) are high points in the history of dynamics (and marvelous reading as well). For the bulk of Descartes on dynamics read Descartes (1983). Chapters 7, 8 and 9 of Barbour (2001) provide a clear picture of the work of the three great precursors of Newton. Dugas (1988) Part I, Chapter IV, surveys the impetus theorists, while Part II, Chapters II, IV and V, gives a brief survey of pre-Newtonian dynamics. Garber (1992) is a study of Descartes' physics in the context of his general philosophy.

CHAPTER 5

The Newtonian synthesis

5.1 NEWTONIAN DYNAMICS

5.1.1 The background to the Principia

In Newton's great work dynamics is presented as being derivable in a systematic way from a small number of fundamental first principles. In this Newton resembles Descartes. But, unlike Descartes, the first principles are not alleged to be derivable a priori from "clear and distinct ideas." They are, rather, painstakingly inferred from the known lower-level generalizations, themselves inferred from observation and experiment, as the best basic principles from which the known phenomena can be derived. And, very much unlike Descartes, and very much in the tradition of Galileo and Huyghens, Newton is extraordinary in his ability to apply the methods of mathematics to the description of particular dynamical situations so that detailed and exact characterizations of the situation can be formulated, and precise predictions made.

Newton's *Principia* appeared only after many years of the careful exploration of dynamics and its applicability to a theory of the world by its author. Most of this earlier work remained unpublished. There seem to be several reasons for this, including many forays by Newton into other fields, such as the invention of the calculus, brilliant experiments on light, including the discovery of the dispersion of white light into colors and basic interference phenomena, and such matters as the invention of the reflecting telescope. Newton's sensitivity to what he took to be inappropriate criticism of early public work may have also contributed to a reluctance on his part to publish.

Fortunately, almost everything Newton ever put on paper remains preserved, and much of the early work on dynamics is now available in print. Particularly important are his early workings in dynamics, collected as the *Waste Book*, and a much later work, indeed a work that immediately preceded the *Principia*, later published as *De Motu*.

Much of Newton's earlier work dealt with problems and issues that occupied others as well. For example, he has much to say about the two problems to which Huyghens made his greatest contributions, collisions and centrifugal force. He often treats the problems in ways that produce results similar to those Huyghens obtained, but by subtly different methods. It is in these methods that we find the germs of Newton's later fundamental principles in dynamics.

In dealing with collisions Newton focusses firstly on the basic facts about inertial motion. Left to themselves, the bodies continue their uniform motions in a straight line. But, in collisions, bodies are not left to themselves; they collide and in doing so "press upon one another." It is this that provides the "motive forces" that change the motions of the colliding objects. Newton makes the fundamental observation that, in a collision of two bodies, the one will induce changes in the motion of the other that are matched by the changes the second induces in the motion of the first. And he accounts for this by arguing that the one object in "pressing" upon the other is "pressed" by the other with a force that is equal but in the opposite direction. From this the results known to Huyghens on the conservation of momentum and the conservation of relative speeds throughout collisions can be derived, as well as the principle that the center of mass of the system remains in uniform motion throughout the collision. It is this work that eventually leads to Newton's Third Law of Motion in the *Principia*. Newton is also very clear about the distinction between elastic collisions, in which energy of motion is also conserved, and inelastic collisions, and he introduces the coefficient of elasticity, k, for the first time.

Newton also takes up centrifugal force in this early work, deriving the same results as Huyghens but by a novel demonstration. He considers a sphere made to bounce around inside a closed polygonal shell, and asks what pressure the shell must exert on the ball at each reflection (and, by application of his principle about forces being equal and opposite, what pressure the ball must exert on the reflecting wall). By going to the limit of a circular confining wall, he obtains the same formula as that of Huyghens. In this work Newton is always clear that the basic principle is that of inertial motion when the object is unimpeded, and that the "motive force" which changes the particle's motion away from what would otherwise be its motion with uniform speed in a straight line is always to be measured strictly in terms of the force's ability to change motion over time.

In this early work on collisions and on constrained circular motion lie the seeds that grew into mature plants in the *Principia*.

The circumstances that finally gave rise to the writing and publication of the *Principia* are themselves a curious chain of events in the history of science. In 1679 Hooke wrote Newton asking Newton's opinion of his (Hooke's) view that the heavenly motions could be viewed as the result of tangential inertial motion and attraction to the center. In his response Newton proposed a way of demonstrating the Earth's rotation by releasing an object from a tower and following its trajectory, imagining it able to move freely through the Earth. In a diagram Newton suggested that the path would be a spiral ending at the Earth's center. To Newton's abashment, Hooke responded by correcting him by pointing out that the path could be, rather, a circular or elliptical orbit around the Earth's center. Newton admitted his mistake, but now had his attention once more drawn to both dynamics and the old problem of the motion of the heavenly bodies.

Meanwhile Halley, Wren and Hooke discussed the issues among themselves. In 1684 Halley asked Newton what trajectory would result from an inverse square law of attraction and Newton replied that it would be an ellipse and that he had calculated this result but couldn't find the calculation when he looked for it. In the same year Newton gave lectures on dynamics (*De Motu Corporum*) at Cambridge and began the text, *Propositiones de Motu*, that immediately preceded the *Principia*. Finally, in a frenzy of work from the end of 1684 through 1687, Newton completed the entire grand work.

5.1.2 The fundamental concepts of Newton's dynamics

Newton begins the formal exposition of his theory with a series of definitions of fundamental concepts.

First he defines a quantity of matter, namely what he, and following him all others, call mass. Here the definition seems quite disappointing, since Newton defines his quantity of matter as the product of density and volume, in a manner that now seems to us both circular and backwards. He clearly intends, however, the mass of an object to be an intrinsic property of the object, and to be distinguished from its mere volume. Most interestingly, Newton then goes on to state that weight is an appropriate measure of mass. Here he means weight at a single place, since Newton is well aware that the weight of an object can vary with its place. He notes that he has performed careful experiments with pendulums to determine with great precision that weight is proportional to what we call inertial mass, thereby improving on Galileo's observations about the equal times of fall taken by bodies independently of their size and constitution.

Next the quantity of motion is defined as the product of the quantity of matter and the velocity. That is, the quantity of motion is momentum.

Newton uses the term "force" (*vis*) in a variety of ways, but he is very clear about the multiple meanings and proceeds to disentangle all of them. *Vis insita* or innate force of matter can also be called *vis inertia*. It is just the propensity of an object to retain its uniform motion in a straight line, that is, to resist the attempt to change speed or motion due to some outside force impressed on the object. Closer in meaning to the standard use of "force" is *impressed force*, which is the action externally imposed on an object to change its state of rest or uniform motion. This is the force that plays such a crucial role in the Second Law of Motion. Centripetal force is that "by which bodies are drawn or impelled towards a point as to a centre." Three further definitions unpack aspects of centripetal force. Its *absolute* quantity is a measure of its efficacy as a cause (such as the active gravitational mass of the Sun). Its *accelerative* quantity is proportional to the velocity which it generates in a given time. Its *motive* quantity is proportional to the motion (that is velocity times mass) that it generates in a given time.

These definitions are followed by a "Scholium," or commentary, that remains justly famous as one of the very most important contributions ever relating to the issues of the "metaphysics" of a scientific theory. Here I shall just outline its most salient claims, leaving the detailed discussion of them to the next chapter, in which Newton's "philosophical" views on space and time and the controversy they generated will be reviewed.

"I do not define time, space, place, and motion," says Newton, "as being known to all." But, even if we take these notions as primitive, more needs to be said. For, Newton claims, we tend to confuse the absolute notions of these terms with their merely relative aspects, that is the "relations they bear to sensible objects."

There is time that is measured by clocks in their repetitions. However, "Absolute, true and mathematical time, of itself, and from its own nature, flows equably without relation to anything external, and by another name is called duration: relative, apparent, and common time, is some sensible and external (whether accurate or unequable) measure of duration by means of motion . . . " Also "Absolute space, in its own nature, without relation to anything external, remains always similar and immovable. Relative space is some movable dimension or measure of the absolute spaces; which our senses determine by its position to bodies." Place is a part of space, whether absolute or merely relative.

Whatever all of this means, one thing is clear. Newton rejects entirely the Cartesian doctrine of relationism, the doctrine that the position of an object is only position relative to some other object and that the motion of an object is only motion relative to some other object. Absolute position is position relative to space itself, not to any sensible object, and, as we shall see, absolute motion is motion not relative to some sensible object chosen as a standard of reference, but to space itself. One of Newton's deepest insights is that his doctrine of absolute place and motion must be accompanied by a doctrine of the absolute rate of the lapse of time if they are to serve his dynamical purposes.

"Absolute motion is the translation of a body from one absolute place into another." In spite of all the tendency toward relationism, beginning with Galileo's remarks on the impossibility of detecting by any mechanical means the motion of a uniformly moving ship and ending in the full-fledged relationism of Descartes, and, as we shall see, of Leibniz, Newton insists that motions are not merely relative but absolute.

When we do astronomy we use absolute time, correcting the measure of time given by cosmic dynamical processes by applying the equation of time. Only such corrected time is suitable for dynamics.

The moments of time have an immutable order in time as a whole, and the parts of space are also immutably placed in the spatial whole. But we have no sensible access to parts of time or parts of space themselves, only to times as measured off by material clocks and places as measured off by material reference objects. So how can we gain any observational access to the absolute temporal and spatial aspects of things? Well, we do have at least some observational access to absolute motions, for "... we may distinguish rest and motion, absolute and relative, one from the other by their properties, causes and effects."

I can generate relative motion in a object by moving the reference object and imposing no impressed force on the object in question. But true motion in an object can be altered only by an impressed force directed on the object itself. Further, the absolute motion of a body reveals itself in effects. A body that is really in rotational motion shows the tendency of all its parts to recede from the center. Merely putting an object into relative rotation by spinning something around it generates no such effects. Put water in a bucket and hang it by a rope. Twist the rope. Let the rope untwist. At first the bucket spins with respect to the water in it. The surface of the water remains flat. Eventually the water picks up the spin and comes to rest with respect to its containing bucket. But then the surface of the water becomes concave as the water, now in real, absolute, rotary motion, tries to recede from the axis of rotation. Thus can true circular motion be shown. So it is with the Earth and the planets in their real orbital motions about the Sun. So much the worse for Descartes' pusillanimous claim that we can think of the Earth as being at rest.

We cannot discern absolute motion directly, since we have no sensory access to the parts of space itself. But, even far from any other matter, we could tell whether or not two globes on the end of a rope were or were not in absolute rotation about the center of the rope. If they are, this will be revealed by tension in the rope; if they are not, there will be no such tension.

Let us leave it at that for the moment. In the following chapter we will return to Newton's arguments here and to their consequences for contemporary and future philosophical inquiries into the nature of space and time.

5.1.3 The laws of motion

The Newtonian scheme is founded upon the famous three axioms or laws of motion. Once again, these are posited by Newton not as somehow obvious to the rational understanding, but, rather, as generalizations inferable from a variety of experimental observations and lower-level generalizations governing these. A perusal of the unpublished work preceding the *Principia*, especially *De Motu*, shows the efforts Newton went through in trying to find the optimal axiomatic basis on which to rest the system.

The First Law is the principle of inertial motion: "Every body continues in its state of rest, or of uniform motion in a right line, unless it is compelled to change that state by forces impressed upon it."

The Second Law identifies change of motion with quantity of motive force impressed: "The change of motion is proportional to the motive force impressed; and is made in the direction of the right line in which that force is impressed." Usually this is read as the familiar statement that "force is equal to mass times acceleration, $F = ma$," but it is probably more accurately read in Newton as saying that the force impressed over a duration of time is equal to the change in the quantity of motion, or momentum, over that time, say $F\,\mathrm{d}t = \mathrm{d}(mv)$. Of course, force, velocity and acceleration are perfectly understood by Newton as directed quantities, as his statement of the law makes quite explicit.

The Third Law is the one most original with Newton: "To every action there is always opposed an equal reaction: or, the mutual actions of two bodies upon each other are always equal, and directed to contrary parts." The inspiration behind this law plainly comes from the earlier work on collisions. It is the "force" version of the realization that the change of motion communicated by one colliding object on another is equal and opposite to the change of motion the latter communicates to the former. But now the principle is being extended to all interactions of one body with another, including the hand which pulls on the rope being matched by the rope pulling on the hand, and, most importantly, those "tugs" at a distance responsible for the interaction of the Earth with falling objects and the Moon and of the Sun with the planets.

The laws are followed up by six important immediate consequences, the corollaries, and by a commentary, the "Scholium," on the laws. Corollaries I and II deal with the composition of forces in different directions and the resolution of forces into directional components. Corollary III states the conservation of quantity of motion in a given direction, and Corollary IV that the center of gravity of a system of bodies will continue in uniform motion throughout an interaction.

Corollaries V and VI are of special interest. Corollary V states that, if a system of bodies involves their mutual interaction, the relative motions of the bodies will be independent of any common uniform motion on their part. That is, it affirms Galilean invariance. Newton assumes in his proof that the forces of interaction will not vary because of the inertial motions, which doesn't itself follow from his laws. Corollary VI states that the same will be true even if the bodies are all undergoing common accelerated motions in the same direction. Corollary VI is needed in order to deal with such motions as that of the moons of Jupiter around the planet. Newton will treat the system as if it is at rest, even though he knows perfectly well that the whole Jovian system is in accelerated motion about the Sun. But that motion is adequately approximated by a parallel acceleration of the planet and each of its moons, given their distance from the Sun and proximity to one another.

Corollary V plainly tells us that no dynamical observation will reveal to us which uniformly moving frame is the frame truly at rest in absolute space, a problem that haunts Newton's theory from its very beginnings. Corollary VI, as was later pointed out by J. C. Maxwell, tells us that two worlds alike in all respects, except that the entire content of one world is in uniform motion and that of the other in uniform, parallel, accelerated motion, will look exactly alike as far as any Newtonian dynamical observations can show. That observation, together with the Galilean–Newtonian observation of how gravity accelerates things independently of their mass and composition, will ultimately become a key ingredient in the theory of dynamics plus gravitation that replaces Newtonian theory, general relativity. The results contained in Corollaries V and VI together will ultimately provide the key to a theory that reconstructs Newtonian dynamics and Newtonian gravitational theory in such a way as to avoid both of the conceptual puzzles noted here in the original theory. The puzzles are twofold but of the same nature: Why tolerate a theory that takes absolute rest as a feature of the world when it is empirically undeterminable what state of motion amounts to absolute rest? Why tolerate a theory that distinguishes uniform, parallel, inertial motion from uniform, parallel, accelerated motion of a system, when, on the theory's own account, the components of the system will show exactly the same relative motions with respect to one another?

Newton finishes up his exposition of the laws and their immediate consequences with another of his brilliant commentaries on his own results, the "Scholium to the Laws." First he states that the first two laws had, along with the first two corollaries, been employed by Galileo in his laws of falling bodies and in his study of the motion of projectiles. The three laws together, he says, had been employed by Wren, Wallis and Huyghens in their studies of collisions. (Descartes, along with Hooke, is not someone to whom Newton will willingly grant any credit whatever!) Newton then has interesting things to say about extending the Third Law beyond collisions so that attractions must be considered to obey it as well. Suppose A and B attract one another but unequally. Put a third object, C, between them. Then, if A is attracted to B more than B is to A, the pressure of A on C will be greater than that of B on C. As a consequence the whole system will accelerate without limit in the direction in which B lies, which is "absurd and contrary to the first Law." Newton then describes how he performed experiments with lodestones floated on water to confirm the equal and opposite forces asserted to be the case by the Third Law for attractions.

The remainder of Book I and all of Book II of the *Principia* are devoted to drawing consequences from the Laws of Motion. Book I is devoted to motion without resistance, and Book II to the motion of bodies in resisting media and major work in fluid dynamics. The results are truly astonishing. Two of the more dramatic results are Newton's amazingly simple demonstration that Kepler's equal-area law must be obeyed by the motion of any particle whose motion is governed by a centripetal force (Proposition I of Book I), and the proof that, if a particle is in motion in an ellipse and has that motion governed by a centripetal force directed toward a focus of that ellipse, then the force must be governed by an inverse square law (Proposition XI of Book I). The final resolution of the problem of a dynamics that could lead to Keplerian astronomical orbits has been obtained. In Scholia to Propositions LII and LIII of Book II Newton makes it clear that the Cartesian theory of the planets as drawn about in vortices in the plenum is a non-starter. Neither the Kepler law correlating orbital times to distance from the Sun, nor the law stating that the planetary orbits are elliptical with the Sun at one focus, is compatible with such vortical motions. It is in Book III that the results of Book I are utilized to provide the correct dynamical theory of the heavenly motions.

5.2 NEWTONIAN GRAVITY AND NEWTONIAN COSMOLOGY

The central thrusts of Book III of the *Principia* are the explanation of the heavenly movements in terms of the fundamental dynamics of Book I combined with the single posit of a force between the heavenly bodies that obeys an inverse square law of diminution with distance, the identification of that force with the force of gravity familiar from the terrestrial phenomenon of falling bodies, and the characterization of the fundamental features of the gravitational force. These features are its universality, with all matter attracting all other matter by mutual gravitational interaction, and the deep connection between gravity and quantity of matter, a connection not exhibited by other forces, say those of magnetism. First, the degree to which an object is attracted by another is proportional to the quantity of matter in the attracted object. Second, the power an object has to attract other objects is also proportional to the quantity of matter, the mass, of the attracting object. Here Newton homes in on the fundamental aspects both of cosmic dynamics and of gravity, treating them in a theory that remained unchallenged until the advent of general theory of relativity over two centuries later.

The origins of Newton's bold idea that terrestrial gravitation and the force that held the heavenly objects in their orbits are manifestations of the same phenomenon are still matters shrouded in some obscurity. Certainly Newton in his old age claimed to have had the idea in his famous retreat from Cambridge during a plague, that is in 1665–66. He claims also to have deduced from Kepler's Third Law of planetary motion, and, presumably from the formula for centrifugal force, the inverse square nature of the force at the same time. In other places, enraged at Hooke's claim that he, Hooke, had been shabbily treated by not being given sufficient credit in the *Principia,* Newton refers to conversations with Wren in which he was sure Wren knew the law and claims that in the 1670s he, Newton, had already been doing the requisite calculations that led to the confirmation both that the force holding the Moon in its orbit was ordinary gravitation and that this force diminished with distance as the inverse square.

The calculation, when finally carried out to Newton's satisfaction, compares the acceleration induced in a falling object near the surface of the Earth, that is to say, one Earth radius away from the Earth's center, with the acceleration the Moon suffers in its orbit at the Moon's distance from the Earth's center. Early calculations disappointed Newton in not seeming to obey the posited law, but these discrepancies were later seen to be due to a poor estimate of the Earth's radius, which was corrected by later geodetic measurement.

Newton was also deeply concerned with a problem that needed to be solved before his calculation could be put on a sure basis. The Moon is attracted by the Earth. With the Moon sufficiently far away, we can think of all the matter of the Earth as concentrated at its central point. But that clearly won't do if we are thinking of the falling object near the surface of the Earth as being attracted, centripetally, by each bit of matter in the Earth, which is Newton's mature view. In a brilliant theorem in the *Principia* Newton is able to show that one can substitute for such an attraction the attraction that would result from the entire quantity of matter of a spherical Earth concentrated at its central point, even if the object attracted is close to the sphere, so long as it is external to the sphere of attracting matter.

That the force governing the motion of the planets around the Sun, and of the moons of the planets around their planets, is governed by an inverse square law follows from their elliptical orbits with the governing body at one focus, the assumption that the motion is generated purely by inertia and a centripetal force directed at the governing body, and Newton's famous theorem from Book I that in such a case the force must obey an inverse square law. That the force is gravity is inferred from the fact that the acceleration of the Moon around the Earth is exactly governed by a force that diminishes as the inverse square from the force of gravity at the Earth's surface. In drawing the inferences here, Newton refers to four "Rules of Reasoning" that he has presented in a prologue to Book III. We shall refrain from stating those or commenting on Newton's methodological claims here, reserving that for the next chapter.

That the force of gravity felt by an object is proportional to its quantity of matter, its mass, is argued for both from terrestrial experiments and from astronomical observations. First Newton has performed an elegant series of experiments with pendulums composed of different materials and of different masses, but designed to have the same air resistance, in which he demonstrates the exact proportionality of weight to mass. Further, the fact that the Jovian moons, say, follow exactly the Keplerian laws shows that, in their case as well, the acceleration induced on them by the Sun must be the same as that which the Sun induces in Jupiter itself, for otherwise noticable eccentricities in the orbits would be induced. That acceleration must, therefore, be independent of anything having to do with the moons' sizes or their composition.

Newton then argues (Proposition VII), "That there is a power of gravity pertaining to all bodies, proportional to the several quantities of matter which they contain." In modern terminology, active gravitational mass, like passive gravitational mass, is exactly proportional to inertial mass.

That the active gravitational mass of a body must be proportional to its inertial mass follows from the fact that passive gravitational mass is so proportional combined with the Third Law of Motion. Let A be gravitationally attracted by B. The pull on A will be proportional to A's mass (as passive gravitational mass). But, by the Third Law, B must be gravitationally pulled by A to the same degree. There must then be a force on B generated by A, which force is also proportional to A's inertial mass, which is in turn now shown to be equal to A's active gravitational mass (or, in modern terms, gravitational "charge").

Finally, we come to the universality of gravitation. First we note that gravity is not just exerted by a heavenly body on its own satellites. For Jupiter's attraction to Saturn noticably perturbs the orbits of those planets when they are in conjunction. Further, the Sun perturbs the motion of the Moon; and both Moon and Sun gravitationally act on the parts of the Earth separately, as is shown by the phenomenon of the ocean tides. Also, Newton has shown that objects are attracted to the spherical mass of the Earth as if all of that mass were concentrated at the Earth's central point. In showing this he assumed that the object is attracted with an inverse-square gravitational force to each part of the Earth, that is, to each ordinary bit of matter that makes up the Earth. And the results of his proof from that hypothesis are confirmed in experiment. Hence we have every reason to believe that gravity is not some force especially generated by the Sun and planets due to their special nature as cosmic objects. It is, rather, a force which every object in the Universe impresses on every other object. And it is a force that draws any two objects together with a magnitude proportional to both of their inertial masses and inversely proportional to the square of their spatial separation.

Newton is aware that a critic will reply that no such force is seen to draw together ordinary objects in our experience, the way, for example, lodestones are drawn together. That is simply because, says Newton, "... the gravitation toward them [ordinary objects] must be far less than to fall under the observation of our senses." Here, of course, Newton is on better grounds than Copernicus was when he explained away the apparent absence of stellar parallax by invoking the great distance of the stars. For in Newton's case we have the weight of the object as a measure of its gravitational tug toward the entire Earth, indicating just how very small the gravitational attraction between two ordinary middle-sized objects must be.

Before concluding, we might mention one additional bit of Book III, one of the most curious in the entire volume. Newton suggests "Hypothesis I: That the center of the system of the world is immovable." He says, as support, that "This is acknowledged by all, while some contend that the Earth, others that the Sun, is fixed in the center." From this "hypothesis" he derives Proposition XI: "That the common centre of gravity of the Earth, the Sun, and all the planets is immovable."

Why? Well, according to Corollary IV to the Laws this center is either at rest or in uniform motion in a straight line; but, if the center moved, the center of the world would move also, which would contradict the proposed hypothesis.

First note the appearance of archaic cosmology here. Despite Bruno and Descartes, is Newton still thinking of the Sun and solar system as very special indeed and central to the cosmos? Probably not, since in other related passages he speaks of the "space of the planetary heavens" when dealing with this issue. Still, his argument here is extraordinary. The sole ground offered for the hypothesis is that it is acknowledged by all! The Ptolemaist has the Earth at the center of the planetary system and at rest. The Copernican has the Sun at the center and at rest. So all agree, the central point must be motionless. As Newton was surely aware, his "Scholium to the Definitions" of the *Principia* told us how to empirically detect absolute acceleration, but not how to determine which smoothly moving reference frame was truly at rest in absolute space. Here is his quite disappointing attempt to fill in the gap. As far as dynamics is concerned, the center of mass of the planetary system must be either at rest or in inertial motion. Assume the hypothesis and it is at rest, finally picking out true rest relative to space itself. We will return to this issue in the next chapter.

SUGGESTED READING

Newton (1947) is essential reading, especially the "Definition," "Axioms or Laws of Motion," and the beginning sections of "Book III, The System of the World." Chapter 10 of Barbour (2001) is clear and concise. Dugas (1988), Part II, Chapter VI, is also useful. Westfall (1980) is the classic biography of Newton. The "Bibliographic Essay" at the end of the volume is invaluable. See also Herival (1965) on the background to the *Principia*.

CHAPTER 6

Philosophical aspects of the Newtonian synthesis

Newton's masterful achievement was constructed under the influence of much previous philosophical discussion and controversy that went beyond the limits of scientific debate narrowly construed. Much that Newton says in the *Principia* also ranges beyond the confines of experimental, or even theoretical, science and passes into the realm of what we usually think of as philosophy. Newton's work gave rise, possibly more than any other work of science past or future, excepting just possibly the work of Darwin and Einstein, to vigorous philosophical as well as scientific discussion. Let us look at some of the philosophical issues behind, within and ensuing from Newton's work.

It is convenient to group the discussions into three broad categories. First, there is the "metaphysical" debate over the nature of space, time and motion. Next there is the debate over what can be properly construed as a scientific explanation of some phenomenon. Lastly, there is the controversy over what the appropriate rules are by which scientific hypotheses are to be credited with having reasonable warrant for our belief. We will discuss these three broad topics in turn.

6.1 THE METAPHYSICS OF SPACE, TIME AND MOTION

There are passages in Aristotle that some read as an anticipation of the doctrine about space and time called relationism, such as when he speaks of the place of an object in terms of the matter surrounding it or talks of time as the "measure of motion." But the full-fledged doctrine of relationism is a product of the scientific revolution. The doctrine is first explicitly stated by Descartes in his later work, is accepted by Huyghens, and is worked out in great detail by Leibniz. In one of the most curious episodes in the history of scientific and philosophical thought, Newton, who all along was philosophically predisposed against relationism, changes the whole character of the metaphysical debate about the nature of space and time by offering an allegedly scientific, almost an "experimental," refutation of the relationist's claims.

In the ancient tradition there is a sense in which there is no real debate going on about the absolute or relational notion of motion. Motion is taken to be a property of an object that is not a merely relative property. An object is either at rest or in motion, and one need not supplement assertions about the state of the object by noting with respect to which reference object that one has in mind the rest or motion is being posited. On the other hand, given the belief that the Earth is at rest at the center of the Universe, the Earth itself, with its cosmic position, provides the standard of rest relative to which objects are adjudged to be at rest or in motion.

The strong impetus toward relationism arose out of the desire, beginning with Copernicus himself, to make the Earth's rotational motion credible. We have seen how Copernicus speaks of earthly things as sharing in the Earth's natural motion. And we have seen how, in trying to back up Copernicanism, Galileo points out how physical experiments fail to distinguish smooth motions in a straight line on the Earth's surface. After all, the ball which is dropped from the mast of the ship, although in motion with respect to the pier, is at rest with respect to the ship itself. No wonder, then, that it drops to the foot of the mast.

Descartes generalized this to the claim that it was nonsensical to speak of an object being at rest or being in motion *simpliciter*. An object could be at rest or in motion only with respect to some other object taken as the reference relative to which rest or motion, and kind of motion, is to be specified. We have seen also how he used the doctrine of the relativity of motion, combined with the suggestion that we usually speak of things as moving when they are in motion with respect to the things contiguous to them, to claim that the Earth, driven in a vortex of the plenum about the Sun, could be properly said to be at rest.

Leibniz gives a worked out account of a metaphysics of space and time that is relationist through and through. A nice presentation of his views can be found in a series of letters he exchanged with Samuel Clarke, a disciple of Newton and defender of Newton's absolutism. Leibniz's views of space and time are actually a portion of a deeper metaphysics on his part about which we can only make the briefest remarks. Partly as a response to the difficulty of imagining a causal relationship between mind and matter, and partly motivated by thoughts about perception and its relation to the world that drove later philosophers to varieties of idealism and phenomenalism, Leibniz posits a world composed solely of spiritual beings and their properties, the monads. These basic constituents have no causal relations to one another. But they experience coherent lives due to a "pre-established harmony" instilled in them by God at their creation, which leads each of them to a programmed existence corresponding to the evolution of each other monad.

But we can understand much of Leibniz's space – time relationism by working in a scheme in which material events occur and material things exist. Events bear temporal relations to one another, they occur before or after one another, and different amounts of time separate their occurrences. Objects existing together at one time bear spatial relations to one another. They are above or below one another; one object can be between two others; they have certain specifiable distances between them. There are, then, two "families" of relations, namely the temporal relations among events and the spatial relations among things.

But there is, according to Leibniz, no "time itself" or "space itself." To imagine such "entities" is as foolish as to imagine that, in a family of people who bear familial relations to one another, there is something that exists as an entity in its own right above and beyond the existing people. No. Only the people exist, although they do bear many familial relations to one another. Similarly, events occur, and they bear temporal relations to one another. Material objects exist (in the misleading version of Leibniz we are dealing with) and they bear spatial relations to one another. But there is no time itself and no space itself that would exist even if no material events occurred and no material objects existed.

Leibniz offers a series of arguments designed to show that the opposite view, say that space existed as a substance in its own right, was manifestly absurd. All of the arguments rest upon the idea that, if time and space existed in their own right, then when things happened in time and where things happened in space would be meaningful questions. But, Leibniz argues, such questions are absurd.

Suppose substantival space exists. Then God could have created the entire material world somewhere other in space than where he put it. But then he would have to act without a "sufficient reason" for putting the material world in one place rather than another. But, according to a fundamental metaphysical principle of Leibniz, nothing happens without sufficient reason. So substantival space cannot exist.

Suppose substantival space exists. Now imagine two possible worlds, alike save that the entire material world occupies different places in space itself in the two worlds. These worlds would be, according to the substantivalist, distinct possible worlds. But they would be alike in every qualitative respect. Here Leibniz is operating under the assumption, of course, that every point in space itself is like every other point, and every direction in space itself is like every other direction, that is, that space is homogeneous and isotropic. But a fundamental Leibnizian principle, allegedly derivable from the Principle of Sufficient Reason, is that if A and B have all qualitative properties alike, then A is the same thing as B (the Identity of Indiscernibles). So the two worlds must be, contrary to substantivalism about space, the same possible world. So substantivalism is wrong.

Finally, were the material world somewhere else in substantival space, it would make no difference whatever in any of our possible empirical experiences of things. But it is nonsense to speak of differences in the world that are totally immune from any observational consequences whatever. So substantival space doesn't exist.

What is time? Time is an order of occurrences, that is, a set of relations among material happenings. What is space? Space is an order of relations holding among material things considered as existing at the same time. Actually it isn't quite that simple, for time and space are orders of *possibilities*. Let us just deal with space. There is empty space in the world (if, that is, you don't believe with Descartes that all space is filled with matter). But how can we speak of empty space, say between here and the Sun, if there is no such thing as space? Well, although there is nothing material between the Sun and the Earth (let us suppose), something *could* be there. To speak of the empty space of the world, even of its geometric properties, is to speak of what the family of spatial relations would be like were there material objects occupying the places of space that are in fact empty. It is these "relations in possibility" that constitute what we are talking about when we talk of empty space, rather than some mysterious space substance waiting to have material stuff coincide in position with it.

But not everyone before Newton or contemporaneous with him was a relationist. Indeed, two clear influences on Newton's thought were Henry More and Isaac Barrow. Both taught at Cambridge and, indeed, Barrow was Newton's direct teacher, who ceded his professorial chair to his more brilliant student.

More, called the "Cambridge Platonist," was an ardent exponent of the doctrine that space existed in its own right. It is "infinite, incorporeal and endowed only with extension." Space, according to More, is "one, simple, immobile, eternal, perfect, independent, existing by itself, incorruptible, necessary, immense, uncreated, uncircumscribed, incomprehensible, omnipresent, incorporeal, permeating and embracing all things, essential being, actual being, pure actuality." Indeed, God is always and everywhere present in space itself. Space is a substance in that it exists in its own right. Even if there were no matter in it, space would still have the same being; and this being is an actuality, not a mere mode of possible relations among material things. The echo of More can be clearly heard again and again in Newton's own philosophical remarks about the nature of space.

There are interesting purely philosophical arguments that can be adduced to support such a substantivalist position against Cartesian – Leibnizian relationism. For example, if there were no such thing as space itself with its own existing actual structure, what would provide the ground for the lawlike behavior of the possible spatial relations among things made so much of by Leibniz? If there were no actual space obeying the laws of geometry, why would it be the case that, whatever material things existed, with whatever spatial relations they had to one another, those relations would be in conformity with the laws of geometry? Arguments of this form are the stock in trade of the philosophical, substantivalist objections to relationism. We shall not pursue them, focussing instead on Newton's novel "scientific" refutation of relationism.

Barrow also believed in an infinite, eternal space that exists before the material world and beyond it. And, he insisted, "... so before the world and together with the world (perhaps beyond the world) time was and is . . . " Sometimes his language takes on a "modal" cast not unlike that of Leibniz, as when he says that time " . . . does not denote an actual existence, but simply a capacity or possibility of permanent existence; just as space indicates the possibility of an intervening magnitude . . . " But, he is insistent, time is not a mere abstraction from motion or change. There is a "flow" of time that is uniform and unchanging. Even if all motion and change in the Universe ceased, time would continue to elapse at its steady rate. We can measure the lapse of time with clocks that are more or less adequate, but no material clock is a perfect measurer of the lapse of time. He is a little vague on how we know the real rate at which time elapses, but suggests that it is through a kind of "congruence" among our various measures that we infer the real rate at which time is elapsing. Once again, Barrow's very words are often discernable in Newton's remarks.

Newton had many things to say about the metaphysics of space and time. In the unpublished work *De Gravitatione* he speaks of absolute place and motion in terms familiar from More. He often has theological things to say about space and time as well, taking the Deity to be eternal and ubiquitous, existing at all times in all places. In one notorious passage he speculates about space being the "sensorium" of the Deity, God's visual field as it were. In other places he puzzles over the metaphysical nature of space, sometimes saying it is like a substance, sometimes thinking of it as an attribute (of the Deity), and in other places saying that it has a nature of its own unlike ordinary substance or accident. But it is not in espousing any such "absolutist" doctrines about space and time, or in rehearsing the usual philosophical arguments for them, that Newton draws our rapt attention. For Newton provides a wholly novel argument in favor of the existence of space as an independent entity over and above material things, and for an absolute measure of the "rate of flow" of time. His argument rests upon bringing to the surface a blatant contradiction latent in Descartes.

Descartes' one fully correct contribution to dynamics was in his version of what became Newton's First Law of Motion. Objects not acted upon by external forces persist in uniform motions in a straight line. But the truth of that law, indeed, the very comprehensibility of what the assertion of the law means, requires that we be able to say what it is to move with constant speed and what it is to move in a straight line. But, if we can choose measures of the lapse of time as we desire, any motion can be regarded as at constant speed or at variable speed as we wish. Constant speed means the same distance covered in the same time, and that implies, if constant speed is not to be arbitrarily asserted or denied of an object, that our measure of sameness of time interval be absolute, or at least invariant up to a linear transformation (that is, a choice of zero point and choice of scale for time intervals). And to say that something moves in a straight line also implies some standard of reference relative to which motion is genuinely straight. Choose some reference frame that is fixed in a material object that moves however you like and any motion can be construed, relative to that chosen frame, as in a straight line or not in a straight line as one chooses. To make the First Law of Motion meaningful requires an absolute standard of lapse of time and an absolute reference frame relative to which uniform straight-line motion is to be counted as genuine uniform straight-line motion.

As Newton argues in the "Scholium to the Definitions" of the *Principia*, we can easily experimentally detect deviation from inertial motion. He chooses his examples from rotation (the non-flatness of the surface of the spinning water in the bucket, the tension on the rope holding together the spheres in rotation about the center of the rope), but examples from linear acceleration would suffice as well. Deviations from uniform, straight-line motion show up in the presence of inertial forces. Therefore uniform straight-line motion is not arbitrarily chosen but fixed by nature and empirically discernable. It is that motion which continues unabated when no forces act on the moving object and it is that motion which generates no inertial forces.

One could put an object into relative acceleration by leaving it alone and applying forces to the reference object relative to which the motion of the test object is to be judged. But such relative acceleration is not absolute acceleration. For an object to be truly accelerated, absolutely accelerated, forces must be applied to the object itself. But, if acceleration is absolute, there must be, Newton believes, absolute place and absolute change of place. For only then could absolute acceleration even be defined.

Finally, absolute motion as revealed by its dynamical effects must be attributed to the Earth along with all the other planets. Only by considering the Earth in truly accelerated motion in its elliptical orbit about the Sun can we understand the need for the mutual attractive force Sun and Earth exert on each other, which serves as the "tether" keeping the Earth from following its otherwise natural inertial, straight-line, motion. So much the worse for Descartes' attempt at keeping on good terms with the Inquisition by using relationism to defend the claim of the Earth being at rest.

We shall be returning to Newton's arguments here in several places, once when we discuss Ernst Mach's critical analysis of dynamics in the nineteenth century and its later variants, and again when we look at twentieth-century reconstructions of dynamics that either utilize various spacetime notions to reformulate the Newtonian theory or provide "philosophical" re-readings of the evidence that Newton relied on to present alternatives to Newton's own theoretical account of the data. Suffice it to say here, though, that Newton is certainly right that any espousal of a dynamical theory that places inertial motion at the very center of its theoretical apparatus cannot be compatible with the kind of spatial and temporal relationism espoused by Descartes and Leibniz. Flat-out relationism as they intended it is not easily reconcilable with the existence of special states of motion that reveal themselves as dynamically distinguished in nature.

Leibniz tried to respond to the Newtonian argument, as it was presented to him by Clarke in their correspondence, but his final response to the Newtonian arguments is quite weak. Leibniz says " . . . I grant there is a difference between an absolute true motion of a body, and a mere relative change in its situation with respect to another body. For when the immediate cause of change is in the body, that body is truly in motion; and then the situation of other bodies, with respect to it, will be changed consequently, though the cause of that change not be in them." But consider a wheel spinning for all eternity in an otherwise empty Universe. There is no relative motion of the wheel with respect to other bodies at all. And there is a sense in which there is no "cause" that sets the wheel in motion. Yet, if Newton's science is right (and Leibniz is not disagreeing with it), the wheel's rotation will show up in its internal stresses. To be sure, each point of the wheel suffers internal forces from the other points of the wheel. These are the forces that simultaneously deviate each point from its inertial motion and hold the wheel together. But Newton will insist that the need for such forces to keep the points of the wheel on their circular orbits must be accounted for in terms of something special about the motion of those points, something kinematically, rather than dynamically, characterized. Otherwise the need for the forces could only receive a circular explanation: "The forces are needed because the points of the wheel are in the kind of motion for which forces are needed." And to characterize what is special about the motion of the points in terms that do not themselves invoke the needed forces can only be to assert that the motion of the points requires those forces because the points of the wheel are deviating from uniform motion in a straight line. And that deviation implies the existence of space as the reference frame relative to which such deviation is real, true, absolute deviation.

It is fascinating to see how Huyghens responds to the Newtonian arguments. Huyghens once said that straight-line motions were all merely relative, but that circular motions had a criterion that identified them, the tension in the rope needed to keep the object in its circular orbit, for example. He later tries to give a relationist account of circular motion in terms of points on a wheel on opposite sides of the axle moving in opposite directions relative to one another. But this won't do, for in a reference frame fixed in the wheel, all the points on the wheel are simply at rest. Huyghens is just assuming the description of the system from the point of view of an inertial reference frame. Furthermore, linear accelerations show up dynamically as well. When the emergency brake is pulled and the train screeches to a halt at the station, it is the coffee in the cups held by the passengers on the train that sloshes out of the cups, not the coffee in the cups held by people on the station platform. Yet, relationistically speaking, the platform is just as much accelerated relative to the train as the train is to the station.

There are, of course, deeply problematic aspects to Newton's account. Although absolute acceleration reveals itself dynamically, absolute place and absolute uniform motion do not. If we accept Leibniz's claim, anticipating later positivism, that it is nonsensical to speak of features of the Universe that have no observational consequences whatsoever, how can we tolerate a theory that posits the existence of both absolute place and absolute uniform motion, but which, on its own terms, declares them as having no empirical import whatever? Newton was clearly aware of the problem of the empirical irrelevance of states of absolute uniform motion. Newton himself points out the important facts in Corollary V to the Laws of Motion. As we saw, the best he can do to repair this gap in his theory is to note the peculiar Hypothesis I that rests on what "all agree to," namely that the center of the solar universe is at rest, and use that hypothesis to then fix the center of mass of the solar system as being at rest. Corollary VI shows that there is an even deeper problem in the Newtonian system, in that even some accelerated motions may have no dynamical effects. Both corollaries rest upon implicit assumptions that go beyond Newton's Laws of Motion, namely the assumptions to the effect that the motions will not change the interactive forces among the particles of the moving systems. Both results will play deep roles in later dynamics. The equivalence of all inertial frames will be fundamental in special relativity and in the reconstruction of Newtonian theory from a spacetime point of view (Galilean or neo-Newtonian spacetime), and the empirical irrelevance of uniform universal acceleration will play its role in the foundations of general relativity and in the spacetime reconstruction of the Newtonian theory of gravity.

6.2 ISSUES CONCERNING EXPLANATION

Philosophers try to characterize the general notion of the nature of a scientific explanation. Usually it is assumed that we can say what it is for something to count as having the right character to be a scientific explanation without paying much attention to the actual contents of some particular science in which the explanations are being offered. That is, it is often assumed that we can make sense of unpacking the *form* of what an explanation must be like while remaining indifferent to the particular *contents* of particular explanations offered in particular scientific theories.

But is that really so? Or is it the case, rather, that our very idea of what sorts of things are to count as explanatory is conditioned by the particular contents of what we take to be our best available explanatory theories? This issue can be nicely illustrated by looking at some of the debates about the nature of scientific explanation that arose out of the Newtonian synthesis in dynamics. But to understand these we must first look at the account of explanation which was most popular among knowledgeable scientists immediately prior to Newton's great work.

The ideals of scientific explanation arising out of Newton's work are best understood in contrast to the ideals of explanation promulgated by Descartes and his followers. For the Cartesians the model of scientific explanation offered was proposed as an alternative to what was, rightly or wrongly, taken by them to be the ideals of explanation of Descartes' predecessors. The Cartesians are constantly contrasting their "modern" notion of scientific explanation with what they consider to be the outworn and foolish ideas of their Aristotelian or "Peripatetic" opponents.

The Aristotelians believed in species of natural motions as well as forced motions. Natural motions consisted in the attempt of objects to return to their natural places in the Universe, such as the motion of falling earthly things, and in the perfect, eternal circular motions of the heavenly bodies. All other motions are forced. For Cartesians natural motions are motions at constant speed in a straight line, inertial motions. All other motions are forced.

For Aristotelians, the world is a place of substance and properties. There are many kinds of properties of things, and properties can inhere in things both in actuality and in mere potentiality. For Cartesians there are only two substances, mind and matter; and only two kinds of general properties, thought and extension. For Aristotelians there are many kinds of changes, comings into being and passings out of being as properties come and go in actuality. Motion, properly so-called, is only one kind of change. These changes are to be accounted for in terms of the four causes. For Cartesians there is only one kind of change in the realm of matter, that is, change describable in terms of the basic notions of time and space alone. For the Cartesians, that is, all material change is motion in the narrower sense of change of spatial place in time.

For Aristotelians, at least in the version of their thinking favored by their Cartesian critics, explanation is often in terms of properties of things that are hidden from our direct observational awareness. Peripatetic physics, the Cartesians say, is incessantly resorting to the attribution of "occult," hidden, qualities to things to explain their behavior. But Cartesian physics denies the reality of such hidden causes, or even the meaningfulness of attributing them to objects. For Cartesians all explanatory features must be "manifest," directly open to our observational awareness.

All explanation of all change, that is of all motion, must take one of two forms. The motion may be natural motion, that is inertial motion, in which case no further explanation of it is needed. If the motion is not inertial, it must deviate from uniform motion in a straight line only because some other motion has directly impinged upon the moved object. A ball is accelerated when another moving ball collides with it. A planet moves in an orbit only because it is dragged along by the vortex of the medium in which it resides. Non-inertial motion is always the result of other, contiguous motion. And the fundamental rule governing this causation of one motion by another is that motion is conserved. The accelerated ball has its motion changed only to the degree that its gain or loss of motion is compensated for by the gain or loss of motion of the ball impacting it.

According to the Cartesians any explanatory account of the world that deviates from the Cartesian pattern must not only fail to be scientifically correct; it must also fail to meet the conditions necessary for something to be a genuine scientific explanation at all. The account Newton gives of the motion of the planets fails in many ways to meet the proper standards for explanation as the Cartesians see it. The Cartesian critique of Newton is twofold, even though, curiously, their two objections are often quite at odds with one another. On the one hand, Newton is often accused by the Cartesians of a kind of reactionary resort to justly condemned outmoded forms of explanation. Newton invokes, say the Cartesians, the infamous occult properties of the Aristotelians. Worse yet, he allows for explanations of motion that do not themselves invoke previous motion as the explanatory element, and he tolerates mysterious influences of objects on one another even when the objects are not contiguous to one another. On the other hand, Newton is often accused by the Cartesians of merely describing the motions of things, and not offering any explanation of their motions at all!

Consider some contrasts between the Newtonian and the Cartesian explanatory schemes. The one element they clearly have in common is the postulation of uniform speed in a straight line as the natural state of motion of things, although, as we have seen, Newton takes the posit of such natural motions to be blatantly inconsistent with Descartes' relationist theory of space and time.

Newton invokes both quantity of matter, namely mass, and force as fundamental concepts in his descriptive scheme. Actually he believes in other primitive qualities of matter as well, such as hardness and impenetrability. There is no obvious way that Newtonian physics can be characterized solely in terms of the kinematic notions of place, time and motion to which the Cartesian is conceptually restricted. Whether these apparently primitive concepts are really needed in the Newtonian theory is something we will return to when we discuss Machian and later reconstructions of Newtonian theory. Neither mass nor force is obviously a "manifest" property, as Cartesians take relative place and motion to be. Furthermore, Newton invokes the notions of absolute place and absolute time interval. Here the basic concepts are purely kinematic in nature, but they are, once again, not manifest as relative place and clock-measured time would be.

For Newton the fundamental explanation of change of motion is force, the force one object exerts upon another, be it a force of contact impulse or the action at a distance of gravitational attraction. Motion need not be accounted for in terms of antecedent motion. Indeed, Newton expresses grave reservations about the correctness of any comprehensive posit of the conservation of all motion, remarking how motion can be generated where none was before and how, by means of friction and like effects, motion can disappear from the world. This is so even though Newton was quite aware of how the conservation of linear momentum for point particles acting on each other by forces followed from his Third Law, and even though, as we shall later discuss, other conservation-of-motion results either follow from Newton's original theory or become deeply integrated into its later formalisms.

Of course, motion need not require, at least in the first instance, an explanatory account in which all causes are taken as acting contiguously in space. We will note below Newton's own preference for explanatory accounts that eschew any genuine action at a distance, but at least on the surface the actions of the heavenly bodies on each other, namely those of gravitational attraction, the actions that govern the whole motion of the cosmos, seem plainly to violate Cartesian precepts that all causes are immediately proximal to their effects.

Newton is very sensitive to the charges laid against him by the Cartesians. On the one hand, he is adamant that his account of motion does not resort to "occult qualities." He notes that, when he speaks of the gravitational attraction one object exerts upon another, he is not positing some hypothetical cause of the motions or changes of motions of objects. He is, rather, merely noting the observable deviations from inertial motions that are induced when objects are in one another's proximity. That deviation is, for both objects, proportional to the product of their inertial masses and inversely proportional to the square of the distance between them, and it is directed along the line connecting the objects. From that the law of gravitational "force" follows, and that is all the law is committed to. If anyone is dealing in the "hidden," Newton says, it is those who propose particular "mechanisms" to account for this mutual gravitational influence bodies have on one another.

There is no simple way to characterize Newton's methodology. On the one hand, his restriction within the main body of the work to the mathematical description of the motions of things summarized in general laws, with its eschewal of the search for hidden mechanisms, makes Newton seem quite the positivist. On the other hand, nothing more infuriates the positivistically minded philosophers of his day, or of later eras, than his postulation of absolute space and absolute time.

Anxious to avoid what he takes to be the pointless and endless controversies that rage between scientists and philosophers, Newton, famously, asserts in the *Principia* that he does not "frame hypotheses" about the nature of the mechanism of gravitational attraction. As we shall see in the next section, he claims that all of the assertions he has made in the Laws of Motion and the Law of Universal Gravitation rest on far firmer grounds than any mere "hypothesis."

Nonetheless, Newton does frame hypotheses – about gravity and about many other things as well. In the "General Scholium" that forms the last section of the *Principia*, in the "Queries" section to his famous work *Opticks*, and elsewhere, Newton makes many proposals about the possible mechanisms that might result in gravitational attraction, that might account for light showing the properties that it displays (many of which were experimentally determined for the first time by Newton himself), and that might explain the various structural and behavioral features of matter of various kinds. His hypotheses about gravity, for example, often have a very Cartesian flavor to them, insofar as they postulate "ethers" that fill the Universe with various fluid properties of pressure and resistance, and whose relation to matter, namely their perhaps having a lower pressure where matter is present, resulting in a "push" that moves matter toward matter, might, possibly, explain the lawlike behavior of gravitational attraction. Such "mechanisms" might also remove the taint of action at a distance from gravity. It is worth noting here that the elements that later function to suggest the replacement of action-at-a-distance theories by theories that propose an ontology of "fields" intermediate between the interacting objects, that is to say the time lapse in inter-particle actions and the violation of conservation of energy that results if one is not very careful in framing an action-at-a-distance theory, play no role in the controversies embroiling Cartesians and Newtonians in Newton's time.

Some of Newton's hypotheses remain only as curiosities in the history of science. Others, such as his particle theory of light, remain, if not really correct, important contributions to the development of later science. Still others, such as his hypothesis expressed in the "Queries" to the *Opticks* that there might be other forces alongside that of gravity by which matter influences matter, and that these other forces might account for such things as the structure and behavior of materials, are prophetic insights into what became large components of the future growth of scientific understanding.

In any case, though, Newton is always careful to distinguish what he is guessing at or speculating at, that is, what he is "hypothesizing," from that which he thinks he has established by experiment, observation and the kind of legitimate inferences from these upon which he thinks the core lawlike assertions of the *Principia* are based.

Philosophically the most important thing to notice about this whole debate is the way in which scientists and philosophers become committed to a doctrine about the very nature of what a scientific explanation is depending on which particular theories about that nature they hold at the time. For Cartesians, what they called "mechanical" explanations were constitutive of what any scientific explanation had to be. Any "explanation" that violated their precepts of being framed solely in manifest kinematic terms, relying solely on motion to generate motion, and demanding contiguity of cause and effect, was not explanatory at all. It was either "mere description" without explanatory force, or it was pseudo-explanation resorting to rejected Peripatetic mumbo-jumbo. As we have seen in the case of Newton's account of motion under the influence of gravity, both accusations were made simultaneously.

With the triumph of the Newtonian dynamical scheme, however, came the wholly new idea of what any putative explanation must be like in order that it be a genuine scientific explanation. If an account of a phenomenon did not resort to natural motions being changed by interactive mutual forces among particles, it could not be a genuinely scientific, or, sometimes, a "causal," or, sometimes, a "mechanical," explanation of what was going on.

Just as Newton's science, in not fitting the Cartesian pattern of appropriate explanation but in triumphing scientifically, cast grave doubt upon the very Cartesian demands for the structure of explanation in general, later science, in not easily fitting into a Newtonian pattern, led methodologists to become skeptical of what had become the Newtonian standard of the necessary conditions to be met by any scientific explanation. We will have a little to say about this later when we discuss the critiques of generalized Newtonianism in science by Mach and others in the nineteenth century.

6.3 NEWTON'S "RULES OF REASONING IN PHILOSOPHY"

Newton had framed dynamics in terms of his three fundamental laws of motion and applied dynamics to a theory of the heavenly motions by supplementing the dynamical laws with a law of universal gravitation. But why should we believe in the truth of the Newtonian account?

Newton himself was highly sensitive to criticism and deeply concerned to anticipate what he expected to be angry and vituperative attacks on his masterwork, the *Principia*. First there were the perpetual battles over precedence in discovery endemic to the science of Newton's day and of our own as well. Newton is careful to give generous credit where he thinks it is due: to Galileo on inertia, on the fact that constant force generates equal changes of motion in equal times, and on the fact that the acceleration due to gravity is independent of the size and constitution of the falling object; to Huyghens, Wallis and Wren on the conservation of momentum in collisions; to Huyghens on the magnitude of centrifugal force; and to Boulliau, Wren and others on the inverse-square diminution of the force holding planets to the Sun. Sometimes, though, he is less than generous, failing to note Descartes' first fully correct statement of the inertia law and Descartes' first statement of a principle of the conservation of motion, even if Descartes got the principle wrong. Newton also failed to give Hooke enough credit for being, perhaps, the first person to state correctly that the motion of the heavenly bodies required inertia and centripetal force alone, there being no need for some tangential driving force. Since much of the *Principia* can be considered a sound refutation of everything Descartes said about the structure of the Universe, the less than generous stance toward Descartes can, perhaps, be understood. Since Hooke falsely claimed credit not only for getting elliptical orbits out of an inverse square law, but also for anticipating Newton's invention of the reflecting telescope, Newton's stinginess in granting him credit can also be understood. Hooke's altercation with Newton over the nature of light also played a role, as we shall see, in Newton's framing of his methodological remarks in Book III of the *Principia*.

But it is not quarrels over precedence that most concern Newton. In 1671 Newton presented the results of his wonderful experiments on the refraction and dispersion of light to the Royal Society. These were published along with some of Newton's speculations about the corpuscular composition of light. Immediately afterward Hooke responded with a critical attack offering his own "hypotheses" about the nature of light to contend with those of Newton. The resulting quarrelsomeness so upset Newton that he withdrew from publishing virtually any of his work until finally persuaded to come out with the *Principia* by Halley. Newton was well aware that his views in the *Principia* were likely to start another round of even greater controversy, especially at the hands of defenders of the Cartesian scheme of explanation.

As we have seen, Newton did not cease "hypothesizing," even within the *Principia* itself, where, in the "General Scholium," speculative thoughts about the mechanism of gravity receive their due. But he is careful throughout the work to isolate such "hypotheses" from the far more important work of developing his mathematically formulated laws of dynamics and of gravity and using these to ground the laws governing the motions of the heavenly bodies. He also takes pains in several places to let the reader know that his grounds for believing in the truth of his laws are not the guesswork of hypothesis, but something that he thinks provides a far more secure basis for scientific belief. If the reader accepts these claims, then the core developments of the work will remain immunized from squabbles of the sort that arise when one bit of speculative scientific guesswork is confronted by other "hypotheses" of the same nature.

One thing Newton does not try to do is to show that his laws can be established by some kind of purely rational thought, that is by a-priori reasoning or by Descartes' "clear and distinct ideas." He affirms the role of pure mathematics in his work and the soundness of his reasoning that follows from its use. But he is well aware of the fact that the soundness of the system as a whole is only as sure as the soundness of its first principles.

These, he insists, are derived, not by any mode of pure thought, but by inference from the facts nature presents to our observation and experiment.

In the "Scholium to the Laws" of the *Principia* Newton says "Hitherto I have laid down such principles as have been received by mathematicians, and are confirmed by abundance of experiments." Galileo had, Newton suggests, discovered the Law of Inertia and the Second Law in his experiments on gravity and motion and had derived from them the famous results on the paths of projectiles. Wren, Wallis and Huyghens, Newton goes on, had discovered the truth of the Third Law in their work on collisions. Here Newton realizes that his generalization of that principle beyond collisions and into the realm of attractions is on more dubious experimental ground, and so he offers both deductive reasons why the law must extend to such phenomena and a confirming experimental test using floating magnets.

The laws, then, are supposed by Newton to be established by observation and experiment that is then generalized from particular experiences to all phenomena by what is commonly called inductive reasoning. To be sure, the contemporary philosopher will realize how many pitfalls stand in the way of someone who wants to underpin their beliefs solely on the grounds of the combination of observation and induction. But Newton is surely right in contrasting the support his laws of dynamics receive from quite direct experience projected by universalization with the more tenuous kind of support an hypothesis that involves the widespread positing of "hidden" entities, properties and mechanisms would receive from its indirect confirmation by its ability to predict confirming results at the observational level. Whatever the problems with induction may be, there is a sense in which inductive reasoning can be distinguished from more general "hypothetico-deductive" reasoning, and there is good reason to agree with Newton that his laws of motion receive their support from the narrower, and hence allegedly more secure, kind of inference.

Newton's most self-conscious reflection on methodology, in particular on the grounds for belief in a fundamental physical proposition, comes in an initial prefatory section to Book III of the *Principia* that is entitled "Rules of Reasoning in Philosophy." The material here is plainly intended to provide the basis for the reasoning that will support the inference to the universal law of gravitation. It is the grounds for that law that provides the content of the first part of Book III, and the application of that law in conjunction with the dynamical laws in order to account for the laws describing the heavenly motions that is the bulk of the remaining content of that book.

There are four famous "Rules of Reasoning":

Rule I: We are to admit no more causes of natural things than such as are both true and sufficient to explain their appearances.

Rule II: Therefore to the same natural effects, we must, as far as possible, assign the same causes.

Rule III: The qualities of bodies, which admit neither intensification nor remission of degrees, and which are found to belong to all bodies within the reach of our experiments, are to be esteemed the universal qualities of all bodies whatsoever.

Rule IV: In experimental philosophy we are to look upon propositions inferred by general induction from phenomena as accurately or very nearly true, notwithstanding any contrary hypotheses that may be imagined, till such time as other phenomena occur, by which they may be made more accurate, or liable to exceptions.

It would be a mistake to think of Newton as here proposing some general grand epistemology in the manner, say, of Descartes. He is, rather, adducing just those rules he thinks will appeal to all rational readers as unquestionably sound, and which will be sufficient to allow him to justify his claims to the effect that it is universal gravitation that is sufficient to provide the needed dynamical basis for all the heavenly motions, and to defend those claims from likely "alternative hypotheses" to be flung at him by Cartesian opponents of his work.

Rules I and II are invoked in Proposition IV, the proposition that first associates earthly gravity with a cosmic dynamical force. We can infer from the work of Book I that the cosmic forces are centripetal, for they obey the equal-area law of Kepler. We can infer that this cosmic force diminishes with distance as the inverse square, for the orbits of the heavenly bodies are ellipses with the attracting center at a focus. But measurement of the acceleration of gravity at the surface of the Earth shows that such gravity at the distance of the Moon, having fallen off by the inverse square of distance, will be just the amount of cosmic, centripetal force needed to hold the Moon in its orbit. So the force holding the Moon in its orbit must be just that gravity: "And therefore (by Rules I and II) the force by which the Moon is retained in its orbit is the very same force which we commonly call gravity; for, were gravity another force different from that, then bodies descending to the Earth with the joint impulse of both forces would fall with a double velocity . . . altogether against experience." We need only the amount of the one accelerative force to get the correct acceleration of rock on Earth and the Moon in the heavens, and, since the effect is "the same" in both cases (appropriately modified in magnitude by the inverse square law), the cause of the acceleration must be the same.

In Proposition V it is argued that the similarity in effect of the moons of Jupiter, the moons of Saturn and the planets in their relation to Jupiter, Saturn and the Sun, respectively, to that of the Moon in its relation to the Earth tells us, by Rule II, that it is "no other than a gravitating force" that retains all these other satellites in their orbits as well. This is defended in a "Scholium" to the proposition by reference to Rules I and II, and to Rule IV as well. Presumably the reference to the last rule is to deny the opponent the right to suggest that some other hypothesis could also do justice to the behavior of the satellites other than the Earth's Moon. For in their cases we don't have the argument that backed up gravity as the force used in Proposition IV. But here Rule IV tells us that we need not hesitate in our induction just because of the mere presence of other hypotheses as possible explanations of the phenomena.

Rule III is especially interesting. Its purpose is expressed in an exegesis immediately following the presentation of the rule itself. First it is argued, presumably against Cartesian rationalism and its skepticism of the reliability of the senses, that "all qualities of bodies are known to us by experiments." According to the rule, then, "we are to hold for universal all such as universally agree with experiments." Here quantity of matter (*vis insita*, inertial mass) is likened to such other properties as spatial extension, hardness and impenetrability and mobility. That all bodies have such features, Newton claims, "we gather not from reason, but from sensation."

"Lastly, if it universally appears, by experiments and astronomical observations, that all bodies about the Earth gravitate toward the Earth, and that in proportion to the quantity of matter which they severally contain; that the Moon likewise, according to the quantity of its matter, gravitates toward the Earth; that, on the other hand, our sea gravitates toward the Moon; and all the planets toward one another; and the comets in like manner toward the Sun; we must, in consequence of this rule [Rule III], universally allow that all bodies whatsoever are endowed with a principle of mutual gravitation. For the argument from the appearances concludes with more force for universal gravitation than for their impenetrability; of which, among those in the celestial regions, we have no experiments, nor any manner of observation."

From observation we learn of the irreducible primary properties of matter available to hand for experimentation. By observation we can extend some of our attributions even to the heavens. Then by the universalizing permitted by Rule III, we can finally arrive at the full attribution of the relevant properties to all matter in general. Thus we are able to project our earthly experience into a general description of the heavens as well.

What about the curious qualification "which admit neither intensification nor remission of degrees" in the statement of Rule III? It isn't completely clear what Newton is concerned about here, but perhaps the last sentence of the discussion following the statement of the rule gives us a clue: "Not that I affirm gravity [that is, weight] to be essential to bodies: by their *vis insita* I mean nothing but their inertia. That is immutable. Their gravity is diminished as they recede from the Earth."

Newton is aware of just how subtle the connection is of mass to weight. In the "Definitions" of Book I he told us that we could measure the quantity of matter in a thing by its weight. And in his discussion of gravity he is brilliantly clear on the fact that both the passive and the active gravitational charges of an object must also equal its inertial mass. But the mass is not the weight. The weight is a matter of a *relation* between the object and the Earth that is gravitationally attracting the object. Change the spatial relation of object to Earth and you change its weight. But you do not change its mass, or its intrinsic gravitational charges for that matter. Our "universalizing" of the properties of what is in hand to properties of things everywhere and anywhere must confine itself to those properties intrinsic to the object and not be applied to those which hold of the object only because of its special relations to objects external to it and which may "intensify or diminish" as those relations change.

Of course, Newton has not provided any infallible recipe to tell us which of the properties we experience as universal of things in our experience really are "intrinsic," and which might very well turn out to be in the end merely relational. It was, after all, a great discovery of Newton and his contemporaries that weight was, in fact, not intrinsic but relational. But, as has been said, it would be misleading to think of Newton's rules as proposals for the foundations of epistemology. They are safeguards against polemic and misguided skepticism toward the results of his mathematical physics, especially toward his revelation of the universal law of gravitational attraction and its role in accounting for the heavenly motions.

SUGGESTED READING

The core manifesto of Newton's argument for absolute time and space is the "Scholium to the Definitions" in Newton (1947). Alexander (1956) contains the correspondence between Leibniz and Newton's defender Clarke which expounds all of Leibniz's characterization of relationism and his critique of substantivalism along with Clarke's Newton-inspired responses. Newton's claim to have avoided mere "hypothesis" and his outline of his (self-alleged) method is in Book III "Rules of Reasoning in Philosophy" in Newton (1947).

CHAPTER 7

The history of statics

Our concern in this book is with dynamics, namely the science of motion, its description and its causes. But mechanics traditionally had two branches, dynamics and statics, the latter being the science of the unmoving, that is of how forces can jointly result in unchanging, equilibrium, states of systems. Although we shall deal with statics only in a brisk and cursory manner, we must pay some attention to its historical development, since principles developed in statics played a fundamental role in the foundations of some important approaches to general dynamical theories. Before moving on to the development of dynamics beyond Newton's great synthesis, then, we shall have to spend at least a little time surveying some aspects of the development of statics prior to the eighteenth century.

Two basic areas of investigation constituted the initial exploration of statics in ancient Greece. The major set of problems that gave rise to statics consisted in attempts at describing the general laws governing the equilibrium conditions for simple machines. In particular, the lever and the inclined plane were the characteristic problems tackled. A second branch of statics began with considerations of the static behavior of objects immersed in fluids, consideration of which constituted the first efforts at understanding the statics of fluids, hydrostatics. It is the former problem area, though, that is of most interest to us.

Although a Greek historian of philosophy attributes the first mathematical work on mechanics to one Archytas of Taras, the earliest existing text we have is a *Mechanics* attributed, almost surely wrongly, to Aristotle. It deals with the problem of the lever, or the equivalent problem of the balance. Empirically it was clear that two objects on opposite sides of a straight lever balanced when their distances from the balance point, the fulcrum, were inversely proportional to their weights. Why was this so?

Pseudo-Aristotle offered a kind of "dynamic" explanation of this static result. If the lever moved, the objects would travel in circles. The further from the fulcrum the object was, the larger the circle in which it would travel. In a fixed time, the further object would travel the greater distance and so, in motion of the lever, the velocity of an object's travel upon rotation of the lever about the fulcrum would be proportional to its distance from the fulcrum. Pseudo-Aristotle proposes, then, a basic principle: The weights on a balance are in equilibrium when they are inversely proportional to the velocities they would have in circles upon rotation of the lever. This idea, when later shorn of the notion that circular, rather than rectilinear, motion has anything to do with the issue, and when profoundly refined by paying attention only to the "infinitesimal" motion the objects would have when the lever first begins to rotate, later developed into the "principle of virtual velocities."

A quite different approach to statics is contained in a book attributed, questionably, to Euclid, the *Book of the Balance*. This approach was taken up and developed by Archimedes. Quasi-dynamical notions such as the primitive versions of principles of virtual velocities play no part and the approach may be called "purely static." Here the lever problem is dealt with in a manner familiar from Greek axiomatic geometry. There are statements of postulates whose truth is supposed to be evident to the reader. From these postulates more complex, less "self-evident," results are derived by logical deduction. *The Book of the Balance* assumes that a lever is in equilibrium when equal weights are suspended at equal distances on opposite-sides of the fulcrum, that a weight suspended anywhere along a line perpendicular to the lever exerts the same force of rotation on it, and that a weight suspended vertically from the fulcrum has no effect. From this it is argued that equilibrium is not disturbed if we move a weight a distance away from the fulcrum while moving the same amount of weight toward the fulcrum by the same distance. And from that is derived a special case of a lever whose balancing opposite-side weights are in inverse proportion to their respective distances from the fulcrum.

Archimedes, in the book *Equilibrium of Planes*, offered eight basic axioms of the lever: (1) Equal weights at equal distances give equilibrium; (2) Equal weights at unequal distances give non-equilibrium; (3) If weights are in equilibrium and something is added to one, it goes down; (4) If weights are in equilibrium and something is subtracted from one, it goes up; (5) If equal and similar figures coincide, so do their centers of gravity; (6) The centers of gravity of unequal but similar figures are similarly placed; (7) If weights suspended at given distances are in equilibrium, equivalent magnitudes at the same places will also be so; (8) The center of gravity of a figure that is nowhere concave lies within the figure.

From these axioms he claims to infer a number of theorems such as (1) that weights in equilibrium at equal distances are equal; (2) that unequal weights at equal distances are not in equilibrium; (3) that, if the weight of a is greater than the weight of b, then equilibrium requires the distance of a from the fulcrum to be less that that of b; (4) that, if two weights do not have the same center of gravity, then the center of gravity of their sum is on the line joining the centers of gravity of the two weights; (5) that, if three weights lie on a straight line, are equal and have equal distances between them, then the center of gravity of the whole will be the center of gravity of the central object. Finally, he claims to show that, if the magnitudes are commensurable, then they will be in equilibrium when distances from fulcrums are inversely proportional to weights. This can be extended to incommensurable magnitudes by his "method of exhaustion," in which Archimedes anticipated the methods of the infinitesimal calculus.

In the nineteenth century Archimedes' derivation of the fundamental law of the lever was subjected to intensive criticism by E. Mach. Mach argued that Archimedes' reliance on symmetry considerations to try to convince one that the axioms of the system were "self-evident" was misleading, and that, in best empiricist fashion, all the basic postulates rested upon generalized inference from experience. Further, Mach argued, in relying on facts asserted about centers of gravity (expounded by Archimedes in a work now lost), Archimedes had begged the most interesting questions by assuming the fundamental law of the lever, namely that the force of rotation is a product of the magnitude of the force at a point and the distance of that point from the enter of rotation, in his derivations. In assuming that one can replace a distributed weight on the lever by the entire weight suspended at the center of gravity, all that is to be proven is assumed.

Hero of Alexandria states a result he attributes to Archimedes, arguing that, in the case of a lever with bent arms, the effective force, static moment as it would now be called, was the product of the weight and the horizontal distance of the weight to a vertical line under the fulcrum. He did not, however offer a formal "proof" of this important generalization of the lever law. He tried, but failed, to find a correct solution to the inclined-plane problem, that is, the problem of finding the force needed to hold a given weight stationary on an inclined plane of specified slope. More generally he accounted for the uses of whole classes of simple machines on the basis of the general principle that force was always diminished to the degree that the distance over which it acted was increased. In Hero we find something of a reversion to the methods of pseudo-Aristotle with the invocation of the "principle of the circle" and the anticipations of later principles of virtual work.

Much of Greek statics was preserved and commented upon by Arabic authors. This early work entered into the Medieval Latin tradition through translations from Greek and Arabic works.

The most famous contributions to statics in the Medieval period come from three thirteenth-century works attributed to Jordanus de Nemore. One, *Liber de ponderibus*, is in the vein of pseudo-Aristotle and its authorship and status are quite unclear. Two other works, *Elementa* and *De ratione ponderis*, the latter being an extension and correction of the former and possibly by a commentator on Jordanus rather than Jordanus himself, make significant contributions to statics. Jordanus has the idea of finding the component of a force in a given direction. In particular, in the case of the inclined plane he realizes that the force along the slope is inversely proportional to its "obliqueness," where that is measured by taking the ratio of the length of the slope to the length of a vertical line it intercepts. This is the correct law that the force needed to keep a weight from sliding down the slope is proportional to the sine of the angle made by the slope with the horizontal. Very importantly, Jordanus derives the lever law by the basic posit that a force that is effective enough to lift a weight, w, a given distance, d, can lift a weight kw a distance d/k or a weight w/k a distance kd. Here we have a clear anticipation of the later principle of virtual work that becomes the central pillar of statics. In the *Elementa* the bent lever is done incorrectly, but in the expanded work the correct result is derived from the work principle. It is also applied to get the correct law for the force of a weight along an inclined plane.

The work principle is stated by Stevin in the sixteenth century as "The distance traveled by a force acting is to the distance traveled by the resistance as the power of the resistance is to that of the force acting." Stevin also gets the law of a parallelogram of forces for forces that are perpendicular to one another, the resultant force being the magnitude of the diagonal of the parallelogram. He offers a brilliant derivation of the inclined-plane law from a postulate of the impossibility of perpetual motion (that he apparently got from Cardan, who got it from Leonardo da Vinci).

In the seventeenth century Roberval justified the full decomposition law of the parallelogram of forces by arguments founded on the conditions of equilibrium for angular levers.

The work principle was adopted by such developers of mechanics as Descartes and Wallis. Early in the eighteenth century John Bernoulli took the work principle one step further toward its final exact and full general statement. We imagine a number of forces holding some object in equilibrium. Imagine now a small motion of the object as a translation or rotation. Then the equilibrium condition will be that the sum of the products of the components of the forces in the directions of the small motions where they apply to the object times these small motions must sum to zero.

Finally, the principle is understood in its full generality, as for example when stated by Lagrange: "Powers are in equilibrium when they are inversely proportional to their virtual velocities taken in their own directions." Here we have gone beyond the translational or rotational motion of an object as a whole. The system is composed of components whose motions relative to one another are subject to constraints, say two point particles in the system being held at a fixed separation by a rigid rod. Forces are applied to the components of the system. Equilibrium obtains when the sum of the products of the components of the forces in the directions of any infinitesimal motion allowable by the constraints and the amount of that motion is zero.

Lagrange, dissatisfied with the familiar grounding of the principle of virtual work on a generalization from the law of the lever and the principle of decomposition of forces, offers his own rationale. Here the principle is inferred as a generalization from familiar physical facts about the equilibrium obtained when the pulleys in a system of pulleys supporting weights are joined together by a rope. The rope provides the constraint, the weights the forces, and the infinitesimal motions of the rope at each pulley when the system is infinitesimally moved the "virtual velocities."

The evolution of the principle of virtual work as the key to general static equilibrium is hardly the whole of the early history of statics. Indeed, we shall have occasion to mention other important results, such as those of hydrostatics and the solutions to particularly important problems in statics such as the catenary and the beam, as we survey some of the history of dynamics subsequent to the Newtonian synthesis. This brief excursion into one thread in the development of statics has been traveled on its own, however, since, as we shall later see, it is the principle of virtual work in statics that provides the key to one of the major developmental paths taken by dynamics in the eighteenth century. But before exploring that path, we shall, in the next chapter, sketch another main direction followed in the explosive growth of dynamical understanding subsequent to Newton.

SUGGESTED READING

Part I of Clagett (1959) provides both crucial texts from ancient and medieval statics and a brilliant survey of this history by the editor. Dugas (1988) briefly covers important ancient and medieval contributions to statics in Part I, Chapter I, Section 2 and Chapter III. Work of the sixteenth and early seventeenth century is surveyed in Part II, Chapter I. Lagrange's statics is covered in Part III, Chapter XI, Section 2. Part I, Sections I–V, of Lagrange (1997) is well worth reading and makes clear the elements of statics that will be absorbed into dynamics.

CHAPTER 8

The development of dynamics after Newton

8.1 FROM SPECIAL PROBLEMS AND AD-HOC METHODS TO GENERAL THEORY

Our outline of the development of mechanics has proceeded as if we could tell a story with a single line of development. But our account up to now has been, as we shall see, somewhat misleading. Even prior to the great Newtonian synthesis, other approaches to the solution of the problems of dynamics were simultaneously being explored. Most of these approaches remained fragmentary and partial until the eighteenth century. For that reason we have neglected them, reserving discussion of them until the more extended discussion of how those programs became solidified, generalized and systematized in the later history of dynamics.

At this point, however, it becomes impossible to deal with matters in a strictly chronological manner. In the years following the publication of Newton's *Principia* dynamics followed a number of distinct, although deeply related, patterns of development. It will be essential to deal with each of these in turn, forcing us to go over the same temporal period from several perspectives. This chapter is preliminary to those that follow. In it I will try to lay out something of the problem situation facing the great developers of mechanics and outline the basic structure of the multiple approaches suggested to deal with that array of problems. We may then proceed to explore the several approaches in detail one at a time.

Newton's great work provided a systematic attempt at a general theory of dynamics. As we shall see, Newton's famous three laws do not encompass a full foundation for dynamical theory. But they are the starting point from which such a fully general and complete foundation may begin. Newton also provided a rich body of methods for dealing with the application of dynamics to the heavens. In Book I and Book III the fundamental methods for dealing with the motion of point particles in a central field are developed. This is the idealization appropriate to the treatment of many problems of celestial mechanics, as Newton himself showed in his discussion of how spherical bodies can be treated as though their mass were concentrated at their center. Of course Newton has not said all that there is to say about even that case of the application of dynamics, but he has set the stage for all later developments.

In Book II of the *Principia* Newton takes up a number of other problems. Most of his attention is devoted to the problem of objects moving in a medium that provides frictional resistance to their motion. But, as we shall later note, he also provides some crucial insights into how to begin to attack the problems dealing with the statics and dynamics of fluid bodies themselves.

But many, many important cases of statics and dynamics are left untreated by Newton. Nor is it in the least obvious how to extend Newton's general principles and methods for their application to these important problem cases. What are some of the fundamental problems that provided the impetus for the development of dynamics, a development that began before the publication of Newton's work and continued after its appearance?

One set of problems springs from Newton's great work on celestial dynamics. How is one to deal with the problem of perturbations, where one has to deal with a multiplicity of objects attracted to one another by their mutual gravitational forces? And how, also, is one to deal with the special problem of the detailed motion of the Moon, where tidal forces make the "point-mass" idealization inappropriate?

But other problem categories were noted long before Newton's work. How is one to deal with the motion of solid bodies all of whose internal dimensions remain invariant under the motion, that is to say, with the motion of rigid bodies? What methods exist, for example, for characterizing the motion of a spinning axially symmetrical top?

How is one to deal with bodies that are compounded out of simple bodies in a systematic way? For example, the methods of Newton and his predecessors make short work of the simple pendulum. But what of a compound pendulum in which several masses are distributed along the rigid length of the pendulum rod? How is one to deal with the case of one pendulum suspended from the bob of another?

The dynamics of objects whose motion is subject to constraints provides another set of well-known problems. A ball or a wheel roll frictionlessly on a plane – what are their dynamics? A bead is confined to a tube bent into an arc and the tube is set into rotational motion about some axis. How does the ball behave? Here it was realized quite early that there should be some methods for dealing with such cases that avoid solving the problem by a full determination of forces, in particular that avoid the difficult problem of calculating the forces exerted by the constraining body on the body constrained. Can some methods be found to develop the dynamics of the constrained body that avoid calculating those constraint forces, but still take account of the fact that the constraints delimit the possible motions of the body in question?

Flexible bodies provide another set of problem situations. Here the distances within the body of all its mass points from one another remain invariant, but the body is free to change its shape in space. How can the motion of such bodies be determined? Some of these bodies can be idealized as one-dimensional (the taut string, for example) and others as two-dimensional (the drum head, for example). It had been known since the days of Pythagoras that the motion of some of these flexible bodies showed some wonderful features. The plucked string vibrated with a special frequency that was determined by its nature and by the constraints imposed on it, such as the points at which its motion was suppressed by binding, and the degree of tension applied to it. Furthermore, these vibratory motions could be compounded, so that a string could sound its fundamental tone and its harmonics as well. Similar vibratory phenomena with their characteristic frequencies applied to drum heads. How could dynamics solve the problem of vibration of flexible bodies?

Other flexible bodies presented other problems. In statics there arose the question of the shape of a cable suspended at two points, or of a cable so suspended from which a "deck" was hung (the suspension bridge). What shapes would such cables attain, and how could these be derived from fundamental principles of statics?

Other bodies presented even greater difficulties for dynamics. Bodies in which the parts could change their relative positions made for particularly difficult cases. In statics the practical problem of trying to determine the behavior of a beam suspended at one end and with a weight at the other became a standard problem case that was treated by Galileo himself, among others. Fluids were of particular interest both in statics and in dynamics. What, for example, would be the static equilibrium condition of a body of fluid in empty space acted upon only by the mutual gravity of its point-mass components? How could the static condition of fluids in containers be characterized? When fluids were in motion, how could that motion be determined by the various forces imposed on the fluid from the outside and by the constraints of the constraining walls inside of which it flowed? If some object were immersed in a flowing fluid, what forces would the fluid impose on the object partially obstructing its path? How would these forces depend upon the shape of that object, and how would they depend on the characteristics of the fluid's flow?

Connecting the work on flexible bodies dealing with their vibration with the problems in the dynamics of fluid was the problem of sound. Since very early times it was supposed that sound consisted in some sort of vibration in the fluid atmosphere induced by the vibratory motion of some solid object such as a lute string or drum head. But how were such vibrations in a fluid maintained, and how were they propagated? So began the exploration of the dynamics of a wave in a medium.

Finally, a special class of problems that dealt with issues of minima and maxima was recognized. Consider, for example, the following classic problem. An object is released at a point and descends to a new point vertically below and horizontally separated from where it began. In its fall it is constrained to follow a curved path. What shape should this curve take in order that the time from initial to final point be a minimum?

Once again, it is important to note that a wide array of problems of these sorts was already the common concern of explorers in dynamics even prior to Newton's great synthetic work. Many of these problems are extraordinarily difficult to solve in their full generality. The task of tackling them continued through Newton's time and thereafter. The task continues even to the present day. Of course, as we shall see in detail, the Newtonian breakthrough provided essential ingredients for taking on the solution of these problems, but it is also important to note that ingenious methods for particular solutions and general methods for dealing with whole classes of problems in a fundamental way were discovered before Newton's work. Long after the Newtonian synthesis, these older methods and newly invented ones, which were also more or less independent of Newton's approach to dynamics, continued to play important roles in the development of dynamical theory.

What were some of the devices by which problems were attacked before any general program of dynamics was available? What were the methods that continued to be employed as auxiliary devices even after some general structures for approaching dynamical problems had become available?

From the earliest Greek work on statics and throughout the development of dynamics even to the present, principles of symmetry have played a fundamental role. Reliance upon the spatial symmetry of a situation to equate different forces with one another, or upon the dynamic symmetry of a situation to equate quantities or changes of motion, has always played a role in fixing constraints upon the possible solutions to a problem. In particularly fortunate situations considerations of symmetry may go a long way toward providing the full solution to a problem in statics or dynamics even without invoking general principles such as a law connecting forces to motions and without specific knowledge of the ways in which the forces involved depend upon the configuration of the system.

Both in statics and in dynamics some fundamental rules that could be used to resolve a problem were often posited, rules that later came to be looked upon as derivative from more fundamental general principles. In statics, for example, there were principles to the effect that a static equilibrium could be obtained only in a situation where an infinitesimal change in the situation could not lower the center of gravity of the system as a whole. In dynamics the principle that the center of mass of an isolated system must continue to move with constant speed in a straight line was often invoked.

One group of posited principles in dynamics deserves particular attention. These are the conservation principles. Thought of in terms of some rule to the effect that "motion is neither created nor destroyed," the principles took on different forms. Indeed, ferocious debate often ensued as to which form correctly captured the true essence of the general idea. Descartes first posited a conservation rule in the form of the preservation under collisions of the scalar magnitude of momentum, an incorrect rule in every sense to be sure, but one that led others to more fruitful versions of conservation rules. Huyghens realized the need to take direction of motion into account when positing a momentum-conservation rule. Leibniz was the first to suggest that it is what will later be called kinetic energy that is conserved. Leibniz was also the first to suggest, in the form of his notions of "dead" and "live" force, the need for a notion of potential energy along with kinetic energy if a correct conservation principle for energy is to be found. It was with the discovery of the principle that energy is (sometimes) conserved that the interminable debates over what constitutes the correct measure of "quantity of motion" began. Huyghens was the first to make spectacular use of both principles of conservation in his treatment of the collision problem for two elastic bodies, for which the application of the two rules together completely solves the dynamical problem. It was not until the middle of the eighteenth century that it was realized that an additional conservation rule, namely the conservation of angular momentum, must be added to the others.

The conservation principles are of such fundamental importance that they remain throughout the history of dynamics key foundational principles. We will later look closely at the changing role these principles played as the theory evolved over the years.

Another group of principles often invoked consisted in "solidification" posits. Suppose one wants to determine the shape a cable suspended at two points will assume due to the force of gravity on its parts. Take a short section of the cable. It will have its weight as a downward force and the tension exerted on its two ends by the remainder of the cable as the other forces governing its shape. Posit that while dealing with this short section one may consider the parts of the cable external to it as rigid in shape. Or, again, consider a fluid body with all parts of the fluid attracting all others by gravitational force. What shape will it form? Following Clairaut, pick any tube of fluid starting at the surface of the body formed, going through the body and exiting at another surface location. Now assume that the entire body of fluid outside this tube is frozen into a solid shape and deal with the static equilibrium that must then obtain for the fluid in the tube. By considering all possible tubes through the body and applying the solidification principle in each case, much progress can be made on this difficult problem in statics. The use of this principle of solidification is a first move in the later more general program of applying the basic dynamical principles to any infinitesimal element of a body in order to solve problems in statics or dynamics.

Alongside this family of principles or posits a number of formal or technical devices that became essential elements in the program of generalizing and completing the framework of dynamics were also gradually introduced. One of these consisted in the development of the infinitesimal calculus of Newton and Leibniz, and the application of this major branch of novel mathematics to the solution of dynamical problems. Newton, of course, avoided the explicit use of such methods in the *Principia*, in keeping with his desire to preempt anticipated critical objections to that work. Another technical innovation was the improvement of the devices of analytical geometry to formulate the appropriate descriptive language for the systems whose dynamics were to be explored. Perhaps the most important innovation was the introduction of a single system of orthogonal coordinates to which all positions and motions would be referred. This in large part replaced the older version of analytical dynamics due to Fermat and Descartes, in which the coordinate axes were chosen as particular directions determined by important components of the systems being described.

But how was dynamics to go beyond the invocation of ad-hoc methods, and beyond that portion of a systematic theory already available in Newton's *Principia*, to formulate a general and systematic theory that could do justice to the entire range of cases for which dynamical solutions were desired?

8.2 THREE DEVELOPMENTAL STREAMS

The period immediately following Newton's great synthesis, that is to say the eighteenth century, saw dynamics develop following three distinguishable patterns. That is, three distinct ways were found to develop proposed fundamental postulates for the theory, and to apply those posits in solving concrete problems. The three developmental streams were not, of course, totally independent of one another. They were, after all, three ways of developing "one and the same physical theory." The three modes of developing dynamics share many common primitive and derived concepts, and the three approaches often make reference to developments taking place along the other directions. Here we shall take just a preliminary glance at each of the three modes in which dynamics developed, saving the details of each program for the chapters to follow.

The first developmental stream we might call the "Newtonian," for it is the mode that follows most directly from Newton's work and which would seem the most direct generalization and completion of the program of the *Principia*.

The basic Newtonian ideas are preserved. There is a preferred state of motion, inertial motion, that is self-sustaining. Only changes in that motion require explanation in the form of intervention into the system. Changes from that inertial motion are to be accounted for in terms of "forces" that parts of a system can impose upon one another or that can be imposed upon a system or its parts from the outside.

But this development of the Newtonian ideas goes beyond Newton's use of these notions in two ways. First, there is the realization that one can apply the fundamental Newtonian $F = ma$ even to *infinitesimal* parts of a system. Anticipated by the "solidification" methods noted above, this becomes one of the crucial elements allowing dynamics to be applied to the problem of the motion of continua, for example to fluids in motion.

Second, there is the realization that, as far as rotation goes, a distinct new principle must supplement the Newtonian Second Law. This is the principle that, along with unchanging straight-line uniform inertial motion when no forces are present, there can be unchanging constant angular rotation about a fixed axial direction when no torques exist. And, parallel to $F = ma$, there must be another fundamental equation that equates applied torque to the product of the moment of inertia, taking the place of inertial mass, and the angular acceleration. Or, in other words, a principle of the conservation of momentum must be supplemented by a principle of the conservation of angular momentum. This move beyond the original Newtonian structure is essential for coming to grips fully with the appropriate methods with which to deal with the dynamics of rigid bodies and the dynamics of continua.

The second developmental stream has its origins in a marvelous generalization by James Bernoulli of an important principle of statics, namely the principle of virtual work. Starting from an attempt to find a general method for dealing with the problem of the compound pendulum, James Bernoulli realized that by adding to the usual forces of statics a kind of "negative dynamical force" whose form is taken from Newton's Second Law, that is "$-ma$," a new general dynamical principle can be uncovered. Just as in the development of the Newtonian mode, what is going on here is in large part the realization that, along with forces, moments of forces, forces multiplied by lever arms, must be taken into account in solving dynamical problems. In this mode of development that insight is taken into account by the very fact that the static principle of virtual work was already a principle generalized from the laws of the lever in which the balance of moments was realized to be essential, along with the balance of forces, in order to find static equilibria.

The insights of James Bernoulli were expanded on by d'Alembert (whose name became attached to the general method by history), and later fully generalized by Lagrange. At the hands of Lagrange this developmental stream took on new and elegant life as it became a general and systematic method for dealing with a very important class of dynamical problems, that is those problems where the motion of a system is constrained and where we wish to solve the dynamics without becoming enmeshed in the useless and difficult task of trying to discover the magnitudes and directions of all the forces of constraint.

The third developmental stream has the most curious origin of all. The crucial hint for finding a new fundamental dynamical postulate came, not for the last time as we shall later see, from outside of dynamics. It came from optics, the theory of the propagation of light. In Alexandria, Hero had noted that one could account for the simple law of reflection of light, namely that angles of incidence and reflection were equal, by noting that the path taken by the light from point of origin to that of reception was then shorter than any other path in which the reflecting surface was met. In the sixteenth century Pierre Fermat marvelously generalized this into a principle of least time for the propagation of light from which the Snell–Descartes law of refraction could be derived. It was the genius of Maupertuis to see that an analogous principle, later to be called that of least action, could provide a new fundamental posit for dynamics. His work, made more rigorous by Leonhard Euler, then became the first of a sequence of formulations for dynamics (and later theories) that rested upon principles stating that in some process a relevant integrated quantity needed to achieve an extremal value along the real path taken by the process as compared with the value it would obtain along other conceivable, compared paths.

As we shall see, each developmental path carries with it its own approach to the fundamental concepts and postulates on which dynamics is to be grounded. Each path also provides new technical methods for applying the fundamental principles to the solution of important classes of particular dynamical problems. Yet, of course, the three modes of developing dynamics are by no means entirely independent of one another. Each program avails itself of insights developed in the other approaches. Each approach is concerned as well with the question of the degree to which the other approaches can actually be derived from the foundational posits of the approach in question.

8.3 PHILOSOPHICAL THEMES IN DYNAMICS

The development of the internal content of the theory of dynamics is accompanied by a perpetual philosophical reflection on the theory. Sometimes philosophical preconceptions motivate the development of particular formal programs. Sometimes philosophical preconceptions are used to critique the very legitimacy of a proposed scheme of dynamics. In other cases philosophical conclusions are drawn as consequences from the scheme developed. In any case, what one sees is a perpetual rethinking of a number of very fundamental ideas about just what scientific method ought to be and what the contents of successful science "must" look like. The rethinking is driven by the progressive development of the science itself. What are some philosophical issues that continually interact with the ongoing science?

One set of issues has to do with the nature of scientific explanation. We ask "Why?" questions about the various aspects of nature. But what counts as the appropriate form a legitimate answer to a scientific "Why?" question must take? Is to answer such a question to produce the causes of the phenomenon? Or do we provide explanations merely by embedding the phenomenon in a general, lawlike pattern of similar phenomena? Or can a legitimate explanatory account provide something else entirely? When we explain what occurs now, must we resort only to contemporaneous or past occurrences, or can explanations advert to what will happen in the future of what is to be explained? Is there any legitimate place in science for explanations that are "teleological" or "functional," or that advert to ends or purposes, or must such categories of explanation be rejected in any legitimate scientific account of the world?

Another set of issues is concerned with the legitimacy of the ontologies we might posit in order to scientifically explain the observed phenomena. What sorts of entities and properties may we posit if we are to offer answers to "Why?" questions that are scientifically "respectable"? Must all our explanations be framed in terms of entities and properties that are themselves within the limits of observability, or may we posit in principle unobservable things and features in our explanations as well? Must all our explanations invoke only "local" occurrences, explaining what happens here and now only by adverting to what is continuous spatially and temporally with what is to be explained, or are accounts invoking "action at a distance" legitimate as well? When we explain motions, must we posit only other motions as their causes, or is it legitimate for us to invoke such notions as "force" to explain motion?

Finally, there are issues of how our account of motion and its causes, dynamics, is to be fitted into our overall scientific account of the world. In particular, how is our theory of the causation of motion to be integrated with our theories concerning the substance of the world, that is, with our theories about what sorts of entities and properties of things constitute the stuff of the world and of its changes?

In the next three chapters we shall see how each of the three post-Newtonian developmental streams of dynamics dealt with some of these philosophical concerns. The philosophical issues will also permeate the later discussions of how dynamics continued its evolutionary development after the eighteenth century.

CHAPTER 9

The "Newtonian" approach after Newton

9.1 KINEMATICS, DYNAMICS AND CONSTITUTIVE EQUATIONS

The development of the Newtonian approach required the sorting out of several distinct components needed for dynamically characterizing a system. In many of the approaches to specific problems, all of these constituent elements were used. But it took quite a while for the realization to sink in that a clear understanding of a general method for solving dynamical problems required the separation of the elements into components that presented themselves individually in the overt method of solution.

One component was kinematics. This encompasses the general principles for simply describing the motions of systems. Here the obvious fundamental concepts are those of space and time. Position, velocity and acceleration, as well as the notion of a time interval itself, are fundamental concepts. Geometry, and its generalization to take account of the element of time added to those of space, is the fundamental discipline needed.

One of the earliest great breakthroughs was the application of coordinate methods and algebraic manipulations of those coordinates to describe the kinematics of systems. Here Fermat and Descartes were the great innovators. We have already noted the advances achieved in the early eighteenth century. These were the application of the methods of the calculus, originally due to Newton and Leibniz, to generalize the algebraic methods, and the realization that many problems could be more easily dealt with if the coordinates were all referred to a single orthogonal coordinate system fixed in one state of motion in space, rather than to relevant directions fixed by the components of the system to be described.

Profound advances came about with the ability to relate motions as described in one reference frame to the description of those motions from the point of view of a frame moving with respect to the first. Euler's characterization of the rotations of rigid bodies depends crucially, for example, on the subtle means he developed for relating the descriptions in frames rotating with the moving object to descriptions in some inertial orthogonal reference frame. Fluid motions could be dealt with adequately only when descriptions in frames moving with elements of the fluid could be related to descriptions, once again, in some single fixed inertial frame.

As early as Huyghens' work on collisions, it was seen how important it was to be able to transform the description of a dynamical process from one reference frame to another frame moving with respect to the first. But Huyghens transformed only from one inertial frame to another. Now it was realized that transformations to frames that were not inertial might be essential as well. Over a long period this led to the notion that one could describe dynamics in non-inertial frames by the usual dynamical laws only if one introduced the famous "fictitious" forces (centrifugal force, Coriolis force) that are invoked in dynamics as done in rotating reference frames.

Along with pure kinematics, other insights of a closely related nature were needed as well. In the study of the dynamics of fluids, for example, it was important to invoke the concept of the conservation of matter, so that elements of the fluid lost to one region had to appear somewhere else. From such conservation principles were derived the continuity equations that are needed to supplement the dynamical principles themselves in characterizing the dynamics of a system.

Beyond kinematics there was dynamics. Here the crucial insight was to generalize on Newton's fundamental dynamical law, the Second Law of Motion. The basic concepts are those of force and mass and subtle generalizations of these, and the basic assertion is Newton's Second Law of Motion suitably generalized and extended by another closely related principle. These two principles become the key to all dynamics. One attributes to substances the fundamental property of their resistance to change of linear motion, their inertial mass. Their linear changes of motion, their accelerations, then, are proportional to the applied force and inversely proportional to that inertial mass. The other principle attributes to objects a second sort of inertia, their resistance to change of their rotational motion, that is their moments of inertia. And their changes in rotary motion then, their angular accelerations, are proportional to the applied torques and inversely proportional to the respective moments of inertia.

But what determines what the forces and torques are? Here what later came to be called constitutive equations are fundamental. The earliest of these might be thought of as Newton's great discovery of the inverse square law of universal gravitation. Here the force exerted on a particle depends upon the relative separations and the masses of all the particles under consideration. Hooke's early spring law, taking the restoring force on a stretched spring to be proportional to the amount of stretch, is another early constitutive law. In the slow development of the statics and dynamics needed to deal with continuously distributed matter it became clear that the description of how the mutual forces depended upon the distribution of the matter could be a complex and subtle matter indeed. There was the realization that for the simple hydrostatics of fluids just a single number was required in order to give the force on any unit area in any orientation at a point in the fluid, that is, the pressure at that point, but for more complicated situations whole arrays of numbers might be required. Over a long period of time such notions as the stress tensor relating the strain (stretching) in an object to the forces exerted in various directions due to that stretching developed.

A fundamental realization needed to develop this Newtonian approach was that the basic dynamical laws could be applied to any part of a system under consideration. In particular, they could be applied to "infinitesimal" parts. The decomposition in the imagination of an object into an array of infinitesimal spatial parts, which could then be thought of as generating by their relative arrangements forces exerted upon one another, became the basic method by which Newton's Second Law, and the later similar law for rotational accelerations, could be applied to composite bodies, including fluids and more complex solid continua.

9.2 THE FUNDAMENTAL LAWS OF DYNAMICS

The development of the Newtonian approach was largely due to Euler. Along with the other great explorers of dynamics in the first half of the eighteenth century, he set himself the task of searching for solutions of a wide variety of problems in both dynamics and statics. The behavior of strings and membranes, of rigid bodies, of particles subject to constraints and of fluids were among the problems to which he set himself.

In the middle of the century he realized with great clarity that the application of the Newtonian paradigm of change of velocity being proportional to applied force and inversely proportional to intrinsic inertial mass could go a long way toward providing a general method for solving all of these problems. What was required was that one applied the general rule to each and every part of a composite body, even allowing as components infinitesimal elements. Taking account of the external forces applied to a non-pointlike body along with the forces each element of the body exerted on each other element could replace the usual method of searching for a solution by invoking one or other of the varied principles noted earlier.

It would appear that for some time, a quarter of a century, Euler believed that this one principle would suffice for the solution of all mechanical problems. Work on such problems as the motion of linked rods confined to motion in a plane or the motion of a rigid body constrained to rotate about a fixed point led him to eventually realize that an additional principle was needed as well. In statics we need not only to balance the forces acting at a point, but also to balance the moments of forces, forces multiplied by lever arms, around any point. Similarly in dynamics, Euler realized, we must not only equate forces applied at a point to changes in linear velocity at that point, but also equate moments of forces, torques, at a point to angular accelerations at the point.

Just as we need inertial mass to serve as the proportional constant to relate linear acceleration to the applied force, so also we need an intrinsic feature of the system acted upon by a torque to tell us how much angular acceleration an applied torque will generate. That is to say, we need the concept of "moment of inertia" to serve for angular motion as mass serves for linear acceleration. What is needed, then, is a concept of the moment of mass, that is, mass at a point multiplied by the lever arm separating the point from a fixed reference point.

And we need a law of the conservation of angular momentum to supplement the law of the conservation of linear momentum. That linear momentum is conserved follows from Newton's Third Law of Motion. Of course, that angular momentum is also conserved was not something wholly unknown before Euler. There is, for example, Newton's famous geometric proof that when particles move subject only to central forces their angular momentum is conserved, and the fact that, as a consequence, a brilliantly simple explanation for Kepler's equal-area law of planetary motion was forthcoming simply from the central nature of the force holding the planets in their orbits. Later in the development of mechanics it became standard for textbooks to introduce the conservation of angular momentum as a consequence derivable from Newton's original Third Law in the case where one is dealing with the motion of point masses that interact with one another only by forces directed along the lines connecting the points. But it was Euler who first introduced the law connecting torque with angular acceleration as a fundamental law on a par with the linear law connecting force to linear acceleration, with the two laws serving conjointly to provide a set of foundational principles by which any dynamical problem could be solved.

Much of the development of statics and dynamics throughout the later part of the eighteenth century and the nineteenth century consisted in the ever-expanding ability to use these two basic laws to characterize the complex motions of rigid bodies, the motions of fluids, and, lastly, the statics and dynamics of continuous media in which the forces and torques generated by strain and motion could be very complicated indeed. But, however complex the constitutive equation that expressed the rules by which the forces and torques were generated might be, the conservation laws of linear and angular momentum provided profound constraints on their form.

9.3 PHILOSOPHICAL REFLECTION ON "FORCE"

The notion of "force" fundamental to the Newtonian approach is one that remained a matter of perplexity throughout the development of dynamics. In the Cartesian approach, the approach in which all motions are generated from previously existing motions, "force" (*vis* in Latin) is taken to be some measure of quantity of motion. After all, the power of one moving object to change another object's motion is related to the motion possessed by the causing object. Of course, there was much debate about how to measure that motion-change-inducing power of motion. Was it to be taken as scalar speed times mass as in Descartes? Or as vectorial momentum as in Huyghens and many others? Or was it rather a function of the square of speed as in the early claim of Leibniz that what we now call kinetic energy was the proper measure of the amount of motion? But, in any case, force was just motion that could induce change of motion.

For the Cartesians Newton's force is, of course, a terrible Aristotelian throwback. It is an "occult quality" that objects possess in and of their nature. For example, all objects, merely by virtue of having their inertial mass, also have what would now be called "active gravitational charge." They possess the property of affecting the motion of other objects by exerting a gravitational force upon them.

The Newtonian notion of force as used in dynamics has its roots in statics. Here the notion of a force, say in the form of the weight of an object, needing to be balanced by other forces is essential to the problem of determining equilibrium. And, given that the problem is one of statics, there is no way of identifying it with some quantity of real motion. Again, torques, moments of forces, were essential and implicit in statics since the first exploration of the law of the lever. From Galileo onward there is the importation into the dynamical context of this notion of force.

But the notion seems to many to be quite problematic. Spatial concepts and temporal concepts seem to be plainly essential elements of the description of the world evident to our senses. At least that is true if one is dealing with the relational concepts of relative position, velocity and acceleration, for example. From the very beginnings of the Newtonian synthesis the corresponding absolute notions were, of course, looked upon with some skepticism by many. But, in any case, where do we find, in the world present to our observation, the forces of the world? There is, of course, the felt kinesthetic sensations when we try to hold up a heavy weight, for example, but rarely is that thought to somehow legitimate the importation into the theory of "force" as a primitive term.

And what is the epistemic status of the fundamental law relating force to change in state of motion, the law that changes in quantity of motion are proportional to impressed forces? This was not the only law whose epistemic status was questioned in the eighteenth century. There was continual debate about whether or not it would be appropriate to take many of the basic principles as somehow derivable from considerations of pure reason alone, or whether, instead they were to be taken to be generalizations from empirical experience.

Euler and Daniel Bernoulli both took force to be a primitive concept of the theory. Daniel Bernoulli apparently took the proportionality of change of motion to force to be a contingent truth. Euler seems to have taken it to be a necessary truth. In a manner that is quite prophetic of some future approaches to the foundations of mechanics, d'Alembert denied the status of force as a primitive of the theory. What after all is our sole measure of the amount of accelerating force present? It is the change of motion induced in the object upon which the force acts, or at least that change of motion divided by the object's intrinsic inertial mass. Why not, then, simply countenance the motions and their changes and take force to be simply defined by these? Or better yet, see whether we can reformulate the theory without even invoking that problematic notion. We shall see shortly how d'Alembert tries to do that.

On the other hand it is important to note that the idea of force as perhaps being constitutive of the very essence of matter is in the air as well. What, after all, is matter if not just the centers from which various forces, gravity acting at a distance and those resistive forces acting locally that we call impenetrability, are produced? Could we not, as Boscovich suggested, think of force as the basic element of the matter of the world?

But, whatever one's metaphysics, the basic rules of the Newtonian approach are clear. There is the natural state of motion of things, motion in a straight line at constant speed. This certainly requires the existence of preferential spatial coordinates to which motions are referred and an absolute notion of time interval in order to make coherent sense, although it remains problematic whether or not it requires the full-fledged Newtonian notions of absolute space and absolute time. There are forces in the world. Some act on matter in a spatially contiguous way, such as those of impact or of tension in a cord. Others appear to act at a distance, such as those of gravity, although it remains to be seen whether some deeper account for these that reestablishes a principle that all causation is contiguous might someday be found.

The role of forces is to change states of motion. Objects possess an intrinsic property of resisting having their natural states of motion changed, their inertial mass. Dividing the magnitude of the applied force by this mass gives the change of velocity per unit time, the acceleration, of the object. In a closed system the principle that vectorial momentum will be conserved applies. Any momentum gained by some components of the system will have to be compensated for by loss of momentum elsewhere in the system.

But to deal fully with dynamical problems where issues of rotation are involved, to be able to deal with the dynamics of rotating rigid bodies or of fluids or of more complex continua, additional principles are needed. Moments of forces, torques, must be invoked as primitive elements; and moments of masses, moments of inertia, must be introduced as essential properties of objects. A law paralleling Newton's Second Law must be added, relating changes of angular motion, angular accelerations, to applied torques divided by appropriate moments of inertia. And, just as linear momentum is conserved in a closed system, so must angular momentum be conserved.

Allowing these principles to apply to every element in a complex body, even to the infinitesimal elements into which we imagine it being decomposed, provides, as Euler claims, a universal method for the solution of dynamical problems. Most of the great developments in fluid dynamics and the dynamics of more complex continua during the late eighteenth and nineteenth centuries consisted in working out the details of these principles, principles primarily due to Euler's important generalization of Newton's original scheme.

SUGGESTED READING

The discovery of the essential addition to Newton's dynamics of the conservation of angular momentum was first emphasized by Truesdell in his "Whence the Law of Moment of Momentum?" in Truesdell (1968). See also Chapter I, Section B3, of Szabó (1977). A profound study of "force" in Newton in the context of earlier seventeenth-century mechanics is given by Westfall (1971). Part III, Chapter VI of Dugas (1988) briefly surveys Euler's contributions.

CHAPTER 10

From virtual work to Lagrange's equation

10.1 FROM VIRTUAL WORK TO "D'ALEMBERT'S PRINCIPLE"

The Newtonian approach to dynamics had its origins in Newton's great work. It is by the full generalization of the Second and Third Laws to make them applicable to all parts of any complex system, including infinitesimal parts, and by adding to the linear laws those appropriate to rotation that the full theory is obtained. The driving force behind the discovery of the full methodology was the ongoing program of finding solutions to particular difficult problems of statics and dynamics. It was only by coming to grips with such issues as the shape of a hanging chain, the vibrations of a drumhead, the motion of a rotating rigid body, and the dynamics of fluid flow that the general principles became apparent.

The developmental stream we are now about to explore also has its origins in the attempt to solve particular difficult problem cases in dynamics. But it develops not out of Newton's work, but out of the methods of statics that long predate the *Principia*. The problems attacked are those involving constrained motion. In the rotation of a rigid object, one might think of each point mass making up the object as constrained to maintain a fixed distance from each other point mass making up the rigid body. Or one might try to determine the dynamics of a body confined to some geometric figure, such as a bead constrained to slide on a rigid rod of some shape when some motion is applied to that rod. Or, perhaps, one might be dealing with a wheel or a ball constrained to roll frictionlessly on a plane surface. The key to solving these problems is to find some method by which the forces of constraint need not themselves be calculated. As we shall see, the method developed implicitly goes beyond Newton in the same way as the improved Newtonian approach did, in that the roles of torques and angular momenta are taken into account along with the forces and the linear momenta.

A very simple problem involving constraints, and the problem whose solution opened up the entire developmental stream, is that of the simplest compound pendulum. A rigid massless rod free to swing about a pivot at one end has two point masses attached to it at different distances from the pivot. How does it move in a gravitational field? The simple pendulum with a single mass falls easy prey to Newton's method, but this new, seemingly simple, problem requires more ingenious methods. Huyghens found a solution to the problem that cleverly invoked the principle of conservation of energy. But there is no clear way to generalize his method to handle even slightly more complex cases of compound pendula, or to use them to gain insights into the way to solve constrained-motion problems in general.

The great breakthrough was due to James Bernoulli in 1703. The actual problem he worked on is that of an oscillating compound lever with two arms that are at a fixed angle to one another, the arms having different lengths and different point masses at the end of each arm. But this problem is equivalent to the problem of the simple compound pendulum. The gravitational force on each weight can be decomposed into a component along the axis of the pendulum rod and a component perpendicular to that axis. The forces parallel to the rod "distribute themselves over the whole axis and there lose themselves completely" and can henceforth be ignored. The other components can be thought of as partly generating the actual motions that the masses experience and partly compensating each other in such a manner as to keep the two weights always in a fixed spatial relation to one another. So the actual motion will be that in which the residual force not used up in accelerating one object multiplied by the lever arm of that object from the pivot will be equal in magnitude and opposite in direction to the residual force left over for the other object times its lever arm. This conclusion follows from the demand in statics that moments of forces must balance in order to obtain static equilibrium. In other terms: Take the component of force perpendicular to the pendulum arm for each object. Subtract from it the mass of the object times the acceleration it will receive in that direction due to the actual motion of the pendulum once solved. Multiply that by the lever arm of the mass. Add the two results together and they must sum to zero, allowing one to solve for the actual motion.

Others picked up on James Bernoulli's breakthrough. D'Alembert credits Euler with applying the method to particular problems, and Lagrange later attributes a general principle to J. Herman. But an important attempt to frame a general principle deriving from the method is that of d'Alembert, whose name is usually given, with the usual historical infelicity perhaps, to the present form in which the fundamental rule is stated. As we noted above, d'Alembert is none too happy with the notion of force, so he avoids more than passing reference to that concept in his presentation of the principle. Instead he tries to deal primarily with motion (velocity) and quantity of motion (momentum). (Dugas, in his history of mechanics, calls d'Alembert's principle "perfectly clear," but Truesdell has the extreme opposite opinion!)

D'Alembert proposes that one decompose the motion impressed on the objects constituting a system into motions "such that [if they] had been impressed on the bodies they would remain unchanged" and motions such that "[if they] alone had been impressed on the bodies the system would have remained at rest." The sum of these two motions will be the motion the bodies in the system actually acquire. This is the parallel to James Bernoulli's idea of the tangential force on the pendulum masses being decomposed into gross forces on the object that would govern its motion were it not constrained and the forces that would result in static equilibrium of the system, the net result of these compounded being the force left over to actually accelerate the system.

D'Alembert's version of this approach is a serious generalization of the specific applications that his predecessors had utilized. Nonetheless, it is a hard method to apply except in limited cases.

The principle and methods of its application familiar to us nowadays are due to L. Lagrange. In his justly famous *Analytical Mechanics* he begins with statics. He takes as his basic postulate the principle of virtual work, offering a new justification of it as a generalization of experience with equilibria of weights suspended by ropes from systems of pulleys. He also introduces his very important method for dealing with constraints, which is now known as the method of Lagrangian undetermined multipliers.

He begins his treatment of dynamics with a brief (and arguably misleading) historical sketch. He notes four past proposed "principles" on which to found dynamics. The first is conservation of "living force," or energy, which goes back to Huyghens, Leibniz and John Bernoulli and was put to brilliant use by Daniel Bernoulli in fluid dynamics. The second principle is the conservation of the uniform motion of the center of gravity of an isolated system, attributed to Newton and emphasized by d'Alembert. Next is the principle of the conservation of "areas" or angular momentum, which Lagrange attributes to Euler, Daniel Bernoulli and Patrick d'Arcy. Finally there is the principle, to be explored by us in the chapter following this one, of least action. Here Lagrange correctly attributes the initial discovery of the approach to Maupertuis and its transformation into a rigorous theory to Euler.

Lagrange, however, chooses as his basic principle a rule that derives from James Bernoulli's initial insight and from d'Alembert's generalization of that insight. Following his own presentation of the principle of virtual work as the cornerstone of statics, however, Lagrange is able to present the principle in a manner that is easier to understand and to apply than what had previously been available. Take the components of a system whose components may be connected to one another in their possible motion by some set of constraints. Take the force applied to each component of the system. Deduct from that force the mass of the component times its instantaneous acceleration when the system undergoes some small virtual motion. Then treat the result as a problem of statics. That is to say, multiply each residual force by the virtual displacement in its direction and sum all of these products over all the components to zero. Notation aside, this is what one finds in the beginning chapters of contemporary intermediate textbooks on mechanics under the name of d'Alembert's principle.

Lagrange then goes on to derive each of the four principles he discussed in his historical section from this one general principle. Most interesting of these derivations is the one that shows the principle of least action to be a consequence of the dynamical virtual-work rule. We will have more to say about this in the next chapter.

Since the principle of virtual work in statics is a generalization of the law of the lever, that angular momentum must be conserved along with linear momentum is implied by this adaptation of the principle to dynamics. In the form given to it by Lagrange this principle remains an elegant formalization of the foundations of dynamics.

10.2 FROM D'ALEMBERT'S PRINCIPLE TO LAGRANGE'S EQUATION

Lagrange puts his formulation of the d'Alembert principle to great practical use in his derivation of what is now called Lagrange's equation. This provides a device for greatly simplifying the solution of a dynamical problem where constraints are involved. Suppose the original coordinates and the time are tied together by a constraint equation. An example of this would be a rigid body where the distances between the individual points of the body remain fixed with time, or a bead constrained to slide on a wire where the spatial coordinates of the bead at any time are functionally related by the equation that gives the shape of the wire. Such constraints are now called holonomic. As an example of non-holonomic constraints one might consider a particle confined to a box where the constraints are given by inequalities on the position coordinates of the particle or a wheel rolling on a surface where it is derivatives of the coordinates that are functionally related, rather than the coordinates themselves.

In the holonomic case one can reduce the number of coordinates needed to describe the system. In the case of a rigid object constrained to rotate about a point, for example, three angular coordinates will replace all of the spatial coordinates of all of the point particles in the body. Lagrange shows how to transform the original dynamics, given by d'Alembert's principle and framed in the original coordinates of the system, into a dynamical equation that invokes only the independent coordinates that remain after all of the constraints have been taken into account. In the special case where the imposed forces are given as gradients of a potential energy this takes on a particularly simple form, giving an equation that is introduced to all students of mechanics at the first level above introductory physics. The equation is of enormous practical use. For when it is applicable (the constraints all being holonomic and the forces all derivable as gradients of a potential that depends only on position) one can solve the general dynamical problem simply by representing the kinetic- and potential-energy functions of the system in terms of the independent generalized coordinates and plugging the appropriate functions into Lagrange's equation.

After Lagrange ways have been devised of generalizing the method to take account of some kinds of constraints beyond the holonomic and some kinds of forces beyond those derivable as gradients of potentials that are functions of position alone.

As a practical method, Lagrange's equation is indispensable. From the point of view of foundations of dynamics, however, the truly important matter is its derivation from the principle of virtual work in its dynamical form. This is what constitutes this stream of dynamics as one that is distinguishable from the Newtonian approach. Of course Newton's Second Law, which was derived ultimately from Galileo's observation on falling bodies that it is uniform change of velocity in time that gives the right rule, is crucial to both approaches. And both approaches must take account of the fact that the general dynamical theory must focus not only on the conservation of linear momentum, but also on the conservation of moments of momentum, of angular momentum. But one approach develops by first applying the Newtonian Second Law in its full generality to all parts, including infinitesimal parts, of a system and then invoking the corresponding angular principle as an additional basic posit. The other develops out of James Bernoulli's brilliant insight that dynamical problems involving constraints submitted themselves to an approach that took over wholesale the principle of virtual work from statics, a principle that evolved as a generalization of the law of the lever, making it applicable to dynamics by simply considering minus mass times acceleration as though it were an additional static force.

Both approaches, however, clearly arose out of attempts at dealing with difficult concrete cases. In the case of Euler it is rigid motions and linked rods in a plane that give rise to the new fundamental posits. In the case of James Bernoulli it is the case of the compound pendulum. The next stream of development, though, as we shall see, has quite a different motivational background.

SUGGESTED READING

Part III, Chapter I, Section I and Part III, Chapter IV, Sections 1–6 of Dugas (1988) outline the history of the virtual-work method. Part III, Chapter XI, Sections 1–4 continue the story through Lagrange. Truesdell's "Whence the law of moment of momentum?," Chapter V of Truesdell (1968), is seminal and very clear. Chapter I, Part C of Szabó (1977) covers the history of this approach in detail. Part II, "Dynamics," of Lagrange (1997) surveys Lagrange's interpretation of the history of this approach and gives a very clear presentation of his lucid general method.

CHAPTER II

Extremal principles

II.1 FROM LEAST TIME TO LEAST ACTION

The third developmental stream of dynamics that came to fruition in the eighteenth century is quite unlike the first two streams in significant ways. It did not arise out of repeated attempts at solving particular difficult special cases of dynamical problems. Nor, when it was discovered, was its primary importance its ability to provide new methods for solving such problems. Its importance for the discipline was, rather, more of a "theoretical" kind, providing new and deep insights into the fundamental structure of the theory. Whereas the other two streams of development carried with them philosophical issues already present in the standard controversies over the mode of explanation in dynamics familiar to Newton and his critics, this third developmental stream opened up entirely new controversial issues concerning what could count as a legitimate explanation in science. It was also curious that these developments in dynamics had their origins not in contemplation of mechanical issues, but, rather, in explanatory accounts offered in the theory of light, in optics, dating back to ancient Greece.

The Greek mathematicians had become aware of "minimization" problems quite early. For example, it was well known that the circle was the shortest curve bounding an area of specified size. Indeed, this is one of the "perfections" of the circle taken by Aristotle to account for the cosmic orbits. The use of a minimization principle by the ancient Greeks that eventually fed into the important role such principles play in physical explanations, though, was the proof by Hero that the law of reflection, namely the angle of incidence of the light on the surface of the mirror equaling the angle of reflection, could be derived by postulating that the distance taken by a light beam to go from a point to a point on the mirror and then to a third point of reception was least when the mirror point was such that the distance traveled by the light along the path was less than that which would be traveled for any other point on the mirror taken as intermediary. The proof is from plane geometry and is very simple.

For many centuries various principles maintaining that Nature always took the easiest or simplest way of accomplishing its ends appeared and reappeared. There was, for example, Ockham's claim that legitimate theories never ought to postulate more entities than are necessary for the simplest possible explanation; and there is Newton's Rule of Reasoning IV in the *Principia*, as mentioned earlier.

The breakthrough toward truly useful extremal principles occurs, however, with the work of Fermat in optics in the mid seventeenth century. Fermat is able to show using a minimalization principle that one can deduce the law of refraction, namely that the ratio of the sine of the angle of incidence of the light to the sine of the angle of refraction is a constant given by what would now be called the ratio of the indices of refraction of the two media. This is the law discovered by Snell and by Descartes. This is the law governing the special case, of course, of light going from one homogeneous medium into another where the two media are separated by a flat surface. Assuming that the light travels with constant speed in each medium, and that the ratio of the speed of light in one medium to that in the other is given by the inverse of the ratio of their indices of refraction (so that, for example, the light travels more slowly in water than in air), Fermat is able to demonstrate that the actual path taken is an extremal path in time. That is, it is the path such that there is no first-order difference between the time taken along that path and the time taken along any neighboring path. He claims to have shown that the actual path is the path which is minimal, that is, the path along which the time taken by the light is less than that taken along any other path, and, indeed, this can be shown in this case.

The reaction of the Cartesians to this demonstration was a howl of rage. For one thing Descartes, like Newton, believed that the speed of light was higher in the denser medium and lower in the more rarefied medium, the very opposite of what Fermat assumed. For another thing, it seemed to them that Fermat was introducing final causes back into science. Of course, they were soon to have to deal with Newton introducing occult qualities and action at a distance as well. But Fermat's demonstration was probably their first encounter with a "reactionary" mode of explanation being proposed for progressive, contemporary physics. We will have more to say about this in the next section of this chapter.

Shortly after Fermat's work, Huyghens developed his wave theory of light. Here light was considered to be a wave in a medium, and geometrical optics was considered as the study of the lines normal to the fronts of the waves. Huyghens assumes, with Fermat, that the velocity of light is slower the higher the index of refraction of the medium. He is able to produce a purely geometrical-optical proof that the law of Snell and Descartes follows from the path of light being a true minimum of time of travel relative to any other path. Later demonstrations were able to derive highly general extremal principles for the propagation of light (the extremal paths not necessarily being minima of time of travel) from the basic principles of the wave theory itself.

The brilliant application of the use of extremal principles to dynamics was the joint work of Maupertuis and Euler. Their texts appeared at about the same time in the 1740s, but they had been in communication over the basic ideas for some time prior to their publications. Euler, in a spirit of generosity not often found in this period in the history of science, credited the basic insights to Maupertuis. This was, after all, the time of the acrimonious debate between the followers of Leibniz and the followers of Newton over priority for the invention of the calculus. It was also the period during which John Bernoulli pre-dated some of his research in hydrodynamics in order to diminish the credit due his son Daniel, sending Daniel into a terrible depression that drove him out of scientific research.

There was, indeed, a famous controversy over the origin of what became called the Principle of Least Action due to Maupertuis and Euler. A certain Professor König attributed the discovery to Leibniz, quoting a letter allegedly by Leibniz in which the principle is stated. On being challenged, however, he was never able to produce the original document. A battle of the academics ensued, enlivened by the ignorant intervention of Voltaire, who, in Maupertuis' terms, "poured gall and filth" on Maupertuis' head, and who was, of course, totally unqualified to have anything to say on the matter.

Maupertuis' work on least action is intuitive and vague. Euler's is a triumph of mathematical precision.

Maupertuis' work is, once again, inspired by optics, for it is Fermat's demonstration of the refraction law from a least-time principle that catches his attention. It is important to note that at this time it is still very unclear how to scientifically treat the propagation of light. Although Huyghens' wave theory is well known, the general consensus is to view light as the motion of small particles governed in their travel by the laws of dynamics. Maupertuis takes Fermat to be heading in the right direction by looking for an extremal principle, but as having been misguided in his choice of the propagation time as the quantity to minimize. Maupertuis also is determined to hold to the Descartes–Newton thesis that the velocity of light is higher in the medium of higher index of refraction. How can this be reconciled with a minimalization rule?

Perhaps following suggestive remarks in Leibniz (as opposed to the almost certainly spurious attribution to Leibniz of the explicit idea in König's claimed Leibniz letter), Maupertuis suggests that in any dynamic process it is the total *action* that is minimized. By action he means the quantity of motion in the sense of momentum multiplied by the distance traveled. In the case of a single particle being considered, this reduces to the product of velocity and distance being minimized, since the mass is just an irrelevant constant here.

His application of this principle is not completely unambiguous, however. He considers two cases of collisions: the direct impact of two perfectly inelastic bodies and the direct impact of two perfectly elastic bodies. In these cases he takes as the relevant distance traveled the distance each particle moves in a unit time, so that the action becomes the product of mass and velocity times velocity. Looking at the velocities before and after collision, and adopting the principle that the change in the action so calculated must be a minimum, the result is simply, in both cases, the production of the law of conservation of momentum.

Maupertuis' treatment of the refraction problem is much more important and leads to truly deep results. In this case the distance is taken as the path length traveled by the light, but, contrary to Fermat, the velocity of light in the medium is taken to be proportional to its index of refraction. Use of a geometric argument quite similar to that of Fermat results in the familiar refraction law being derivable from a minimalization (or, better, extremalization) posit.

Euler's more general work on the problem appeared as an appendix to his extremely important work on isoperimetric problems. In that work he invented what is now called the calculus of variations, that is, the systematic method for finding paths that will make some integral along the path take on an extremal value. This is the key method for solving all such minimalization (or maximalization) problems for quantities calculable as generalized sums along paths. Euler's solution holds for particles moving along paths confined to a plane. He adopts action as the appropriate quantity to be extremalized, and, dealing with a single particle, can once again neglect the particle's mass. So it is the integral of velocity over distance that must take on the extremal value.

Euler looks at the value of the action for a single particle calculated along the actual path a particle will follow if it obeys Newton's laws when moving in a plane governed by a central force, and the action that would result had the particle followed any other path for which the energy of the particle retains the same value as that which it has along the actual path. He shows that the actual path is the one with the extremal value. He uses the results he obtained in his calculus of variations, work that resulted in the general formula which, when applied to dynamics, results in the same equation as that derived by Lagrange from the dynamic principle of virtual work, that is from d'Alembert's principle. He is able to show that the radius of curvature of the path derived by demanding that the action be extremal relative to all paths with the same total energy is the same radius of curvature as that derivable directly by using Newton's laws.

The full statement of the general principle appears in Lagrange, who deals with any system of particles moving under their mutual interaction according to forces derivable from a potential. The principle is that the sum of all the actions for each of the particles (each particle's mass times the integral of its velocity from its initial position to its final position, the integral being taken with respect to distance along the path followed by the particle from initial to final configuation) is extremal relative to the same quantity calculated for any nearby motions from the same initial to final configuration, with the nearby motion having the same energy as the actual motion. As we noted, Lagrange goes on to emphasize the fact that both the conservation of energy and the extremalization of action are derivable from the first principle of dynamics he has chosen, namely the d'Alembert principle. Plainly they are derivable from the Newtonian version of first principles (the joint principles of conservation of momentum and of angular momentum) as well.

Later versions of the procedure take the extremalization of the difference of kinetic and potential energy along the path as fundamental. In this variant (Hamilton's principle) one demands that the actual path be the one that extremalizes the integral compared with any nearby alternative path. The restriction to other possible paths in which the system has the same energy is no longer needed.

11.2 LEAST ACTION AND THE ISSUE OF EXPLANATORY LEGITIMACY

The invocation of extremal principles as explanatory devices in optics and dynamics was fraught with controversy.

First, there were the repeated attempts to ground these principles on some a-priori metaphysical or even theological view of the world. For Maupertuis it is self-evident that a deity constructing a world would require it "that Nature, in the production of her effects, always acts in the most simple ways." Even Euler, usually the purest of mathematicians, seems to hold that it is a fundamental metaphysical rule that "all processes in nature obey certain maximum or minimum laws."

Initial doubts about the metaphysical status of the principle, or at least about the possibility of deriving it a priori from some fundamental posits of ultimate simplicity, perhaps traceable back to the omnipotence and benevolence of the deity, arise, naturally, out of the fact that it seems very hard to understand how one could, on such a-priori grounds, justify the fact that it is *action* that is made extreme in dynamical processes and not some other calculable integrated dynamic quantity (say kinetic energy over time). Why should it be the product of momentum and distance which must be taken to be the appropriate quantity whose integrated value is to be stationary relative to that obtained on nearby "virtual" paths?

Maupertuis, brooding about this, criticizes Fermat for minimizing time of travel in the optical case, arguing that, since in refraction light already eschews the shortest-distance path, there is no reason to expect it to take that which minimizes time of transit. The very fact that Fermat's principle is correct for light and Maupertuis' least-action principle is wrong, the former resting on the correct and the latter on the incorrect relation of light velocity to index of refraction, gives us, of course, a very strong reason indeed for distrusting a-prioristic arguments here. If light has no reason to choose the shortest-time path over the shortest-distance path (it certainly does not choose the shortest-distance path according to anyone's view about light velocity in the two media), why should it choose either, reasons Maupertuis. It chooses, he says, "a path which has a very real advantage," that is, the path which minimizes action. Which, of course, it actually does not – but which mechanical systems do! It is hard to think of a clearer case in the history of science of the right conclusion arising for the wrong reason.

Euler shows that the path in dynamics chosen by extremalizing action is identical to that which would be obtained by the use of Newton's laws. At least this is so in the single-particle case. His reason for thinking that it will hold in the case of many particles is, curiously, a-prioristic. "For, since bodies resist every change of their state by reason of their inertia, they yield to the accelerating forces as little as possible, at least, if they are free. It therefore follows that in the actual motion the effect arising from the forces should be less than if the body or bodies were caused to move in any other manner. Although the force of this conclusion does not yet convince one as satisfactory, I do not doubt that it will be possible to justify it with the aid of a sane metaphysics. I leave this task, however, to others who are proficient in metaphysical studies."

It is a fact that dynamical systems can be shown to follow a path that is extremal in the action. But this path is not always a path of minimal action, and, indeed, the path will be in some cases a path of maximal action relative to nearby paths. This weighed against taking a derivation of the principle from theologically related metaphysical considerations too seriously. So also did the fact that it is only with regard to nearby paths with the same energy that the actual path is to be compared. This gave to the derivation a "technical" aspect that also cast doubt on its derivability from metaphysical first principles.

For Lagrange any such metaphysical speculations are of no avail whatever. It is with a sense of both impatience and relief that he shows that the general principle, along with the other early principles he itemized in his historical review, can be taken to be just one consequence among many of posited fundamental dynamical laws, in his case the d'Alembert principle.

Next there was the disturbing aspect of the extremal principles that they seemed to some to reintroduce into physical reasoning the Aristotelian final causes that had been declared, especially by the Cartesians, to be totally out of place in physical explanation. Light goes from point A in one medium to point B on the boundary of the media and then to point C. Fermat, given his assumption about the speeds of the light in each medium, shows this to be the least-time path from A to C. But Claude Clerselier, a Cartesian, won't have this. The light is at B. How does it "know" to go to C? If it were obeying the Fermat rule, only by remembering that it came from A! But such an explanation can never be suitable to physics. Only a Cartesian account in terms of the local forces acting on the light at B will be legitimate.

Fermat's reply is truly charming. He admits to not being in the confidence of Nature's "obscure and hidden ways," but only of having "offered her a small geometric assistance in the matter of refraction, supposing that she has need of it." He has demonstrated a mathematical regularity from which the empirically correct law can be derived and eschews the hunt for metaphysical or physical real causes. One is reminded of Newton's "*hypotheses non fingo*" of some twenty-five years later in a similar reply to a Cartesian attack on the explanatory legitimacy of his gravitational law.

Maupertuis takes a more metaphysical bent in his defense of the legitimacy of extremal explanations in physics. He expresses sympathy for the "repugnance" in which final causes are held by some mathematicians. Why? Because they are risky. The risk is not in the general principle that Nature acts in some simplest or minimalizing way, but in knowing the right quantity to be minimized. After all, didn't Fermat fall into the error of thinking that it was time of travel that light minimized? But such errors are just the result of hasty speculation. Deep reflection will pick out the right measure of that which is to be minimized – the action. In the end "It cannot be doubted that all things are regulated by a Supreme Being who, when he impressed on matter the forces which denote his power, destined it to effect the doings which indicate his wisdom." Not for the first or last time Maupertuis is happy to argue from the appearance of a kind of apparent intentionality in the world (dynamical systems acting "in order" to minimize action) to the existence of a conscious intender directing the play of the material world. In finding the principle of least action at the basis of all dynamics, Maupertuis believes he has plumbed the innermost principle of nature. The fact that it is a rule invoking final causes, and that this to him implies an omniscient and omnipotent designer behind the doings of the natural world, is all to the good.

As we noted, even Euler, in many ways the most "modern" of all the great mathematical physicists of his time, was happy to contemplate the source of a general principle of least action in some "sane" metaphysics, even if he felt that origin of the principle to be beyond his powers of discernment.

D'Alembert is highly skeptical of any attempt at founding the laws of dynamics on any reference to final causes, "that is, according to the intentions that the Author of nature might have formulated in establishing these laws." Interestingly he takes Descartes' attempt to found dynamics on a general conservation principle inferable from the "Creator's wisdom" to be one such attempt, similar to the attempt at finding metaphysics in the extremal principles, The extremal principles, however, seem to bring with them an almost irresistible temptation to infer from them to a consciously formulated plan lying behind the world's dynamics. For d'Alembert the only secure foundation for dynamics is "the laws of equilibrium and of motion," presumably in the form of his version of the dynamically generalized principle of virtual work.

After over a century of such controversy it is hardly any wonder that Lagrange wants, by showing that the extremal principle follows as a consequence of his general formulation of the d'Alembert principle, to take the extremal rule as a mere *consequence* of the truly fundamental dynamical principle. This will, he hopes, remove from science any speculation over final causes or over the Creator's plan for the Universe. It will also eliminate the need for any anticipations or memories on the part of the components of dynamical systems.

SUGGESTED READING

Part III, Chapter V of Dugas (1988) surveys the early history of extremal principles, and Chapter XI, Section 6 of the same part goes over Lagrange's derivation of extremal results from virtual work. Chapter II, Sections B, C and D of Szabó (1977) cover the history of extremal principles from the seventeenth until the nineteenth century. Part II, Section III, Subsection VI of Lagrange (1997) is Lagrange's derivation of least action from virtual work. Lanczos (1970) gives a clear survey of virtual-work approaches, extremal principles and their inter-relations.

CHAPTER 12

Some philosophical reflections on explanation and theory

At this point it will be useful to interrupt our exposition of the historical development of dynamics in order to make some brief retrospective observations on how the development of the theory up to the later part of the eighteenth century carried with it an ongoing and evolving debate about the very nature of scientific explanation and scientific theory. What are the legitimate forms an explanatory account can take in science? What are the legitimate concepts that may be employed in such explanations? What are the fundamental posits of our theory, and what are the legitimate grounds by which we may justify our beliefs in the fundamental posits of our scientific account of the world?

The Aristotelian account of dynamics, and Aristotle's related account of cosmology, employed explanatory notions adopted from our pre-scientific, "intuitive" ways of answering "Why?" questions about the world around us. The ultimate origin of our employment of such explanatory structures is a worthy topic for exploration, but one we will not be able to embark on here. The notion of efficient cause presumably comes from our everyday experience, primarily, one imagines, experience of things pushing and pulling each other around. This "everyday dynamics" as it appears in our daily experience may very well be the source of our idea that explanations are to be given in terms of something like Aristotelian efficient causes. The idea of an explanation given in terms of final causes may have its origin in the fact that so much of our activity as agents is accounted for by means–ends explanations in which motives, purposes and goals play such a significant role. Again the idea that various components of living beings, their organs, all have their specific functions or roles, and that their existence must somehow be accounted for in terms of those roles, long predates science properly so-called.

Applying the notion of efficient cause to dynamics and the cosmos, Aristotle comes up with his clockwork mechanism driven by the *primum mobile* for the cosmos and with the various notions of push–pull dynamics for forced sublunary motions. Applying the notion of final cause, we have the "natural" orbits of the heavenly bodies and the motion of objects (such as freely falling rocks) seeking their proper locational homes in an Aristotelian Universe. Along with that notion of final cause we also have the idea of rest as a natural state of things, assuming they are not already "out of place" in the cosmic sense. It is that natural state that requires any forced motion to have a sustaining force and leads, of course, to the notorious problems encountered by Aristotle in dealing with projectile motion.

Aristotle is certainly prepared to infer the existence of explanatory elements that are not immediately observable. Consider, for example, the postulated spheres bearing the heavenly bodies and the prime mover. By and large, however, most Aristotelian accounts are framed in terms of concepts dealing with that which is available to perceptual experience. There is all the talk of that which exists only potentially coming to actuality, of course, but explanations in terms of hidden properties, the Cartesian's hated "occult qualities" seem more a matter of later putative scientific thought and are not to be found in Aristotle himself.

Although Aristotle says little about what he takes the epistemic status of his fundamental principles to be in dynamics, in his general philosophy of science, primarily in the *Posterior Analytics*, he advocates a position in which one takes the fundamental explanatory principles to be ultimately derivable on rational grounds. With our limited understanding, however, we must often be content with a kind of upward inference of the explanatory principles from the data of experience, a sort of inductive reasoning to the basic explanatory rules.

With the coming of the Scientific Revolution more than one general program can be discerned.

There is the course of thought that one can follow from Galileo through Huyghens to Newton and his Newtonian successors. Here there is, again, a natural dynamical state which in itself needs no explanation. But now it has nothing to do with natural place in the Universe. Nor is it "rest." It is a natural state of motion, inertial motion. Coming out of impetus theory, such natural motion is, first, in Galileo horizontal motion on the surface of the Earth sharing in the Earth's own motion, so as to obviate the classical objections to the Earth being in rotation. As such it is a kind of "circular inertia." Only later is the matter refined by Descartes into motion with constant speed in any straight line.

In the Galileo–Huyghens–Newton tradition there develops the notion of "force." Taken over from statics, and finding its paradigm example in weight, it plays the role of a fundamental primitive in dynamics. It is change of motion that is to be accounted for, and this is accounted for by imposed force. The development proceeds from Galileo on the law of falling bodies, through Huyghens on centrifugal force, into Newton's Second Law, and finally into the generalization of that law and its supplementation by a law for torques inducing angular accelerations in Euler.

The other strand of thought that contends with this approach is that due to Descartes. Here the epistemic claim is straightforward. The basic principles of science are to be derived by pure reason, from "clear and distinct ideas" in Descartes' terminology. The model for dynamics is the alleged derivability of geometry from self-evidently true first axioms.

For the Cartesian all legitimate concepts must deal with that which is on the surface of observation. Indeed, insofar as physics is concerned all the genuine concepts must be those of the spatial and temporal nature of the world. Position, velocity and acceleration together with such notions as contiguity and shape constitute the legitimate vocabulary of fundamental physics. Further, all such notions are relational notions. Positions are positions of material things relative to one another; velocities are relative velocities. Without Aristotle's cosmos of a resting central Earth and bounding stellar sphere, with a space that is homogeneous, isotropic and infinite, there is no place for absolute notions of place or rest.

For all the claims about the a-priori nature of science, it is quite clear that it is the observable facts about collisions that give rise to many of the most important Cartesian ideas. Inertial motion is taken as the natural state of things requiring no external explanation. Here, finally, the principle is stated in its fully general and fully correct form: motion with uniform speed in any straight line persists without change unless interfered with. Interference takes the form of contact of one moving object with another. The "force" that leads to change of a state of motion is the motion of the object met; and the basic principle, the law governing "force" and change of motion, is that motion is never gained or lost, it is conserved. In Descartes the principle is muddled by his poor choice of conserved quantity, namely scalar speed, but in Huyghens (vectorial momentum) and Leibniz (kinetic energy) one has the dawning of the true conservation principles.

In the Cartesian program all causes are efficient causes and there is no place for a final cause in any of its senses. Everything, including the motion of the cosmic objects embedded in the swirling plenum fluid filling all space, is governed by the push and pull of contiguous moving objects under the rule of the conservation of motion.

So the laws of nature are a priori. They deal only in concepts of the perceivable. And all causation is efficient causation and local causation. Much of this Cartesian dogma is accepted by Leibniz (but with his own curious metaphysical underpinning in the form of the idealist monadology and with aspects of physics that are much more congenial to the Newtonian picture such as Leibnizian "dead force," or potential energy as we would now call it, and by Huyghens, despite the latter's major contribution to the demise of the Cartesian system in his generalizing on Galileo's notion of force that feeds into the full-fledged Newtonian notion.

With Newton nearly all of this Cartesian scientific edifice comes crashing to the ground, taking with it most of the Cartesian understanding of scientific methodology as well. In Newton there is no pretence at having derived the fundamental dynamical laws from any self-evident first principles (though there are many assertions made in some of the commentaries on these laws that seem pretty dogmatic and "metaphysical" in their nature). The laws are inferred, rather, as generalizations from earlier generalizations from nature (the Second Law from Galileo's law of falling bodies, for example) or from evident facts of nature (the Third Law from the non-existence of self-accelerating systems and from the familiar facts about collisions, for example). In the case of the law of universal gravitation Newton is, of course, explicit in his "Rules" about the inductive grounding for belief in the principles.

Force, at least that "*vis*" which is the motive force responsible for changes in motion, follows the Galileo–Huyghens tradition, taking over and generalizing the notion of force, primarily as weight, from statics. The causes of motion are not merely existing motions that can be transmitted from one object to another, although Newton admits that such motions, if hidden, might play a role even when they seem not to be present, a line taken up much later by Hertz as we shall see.

Worse yet for the Cartesian picture, the solution to the problem of the motion of the cosmos requires the universal force of gravitation. Here all matter has attributed to it an intrinsic propensity to attract all other matter, leading the Cartesians to complain vociferously about the reintroduction into physics of the "occult qualities" they had hoped to expunge from it forever. To add insult to injury the force is treated as if it acted at a distance, violating the Cartesian demand for all efficient cause to be local and contiguous. Here, once again, of course, Newton is willing to admit that it might be possible that this action at a distance is underlain by some hidden mechanism that is local in its action – but is unwilling to frame speculative hypotheses about what such a mechanism might be. Such is the hold of the Cartesian ideal even on Newton himself.

Add to all of this Newton's trenchant demonstration that Descartes' spatial relationism was directly incompatible with his principle of inertia. A preferential, non-forced, kind of motion with constant speed in a straight line makes no sense unless one has a notion of what constancy of speed is that is independent of particular arbitrary markings off of equal time intervals by arbitrary clocks. The notion of a straight line invoked also requires that this line be straight relative to something other than some randomly chosen and randomly moving spatial reference frame. Inertial motion must be an absolute, non-relational notion if it is to do its dynamical work. For Newton the solution is, of course, absolute space and absolute time themselves. Here the Cartesian ideal of a theory that uses only the surface observables is under dire threat.

As if enough destruction hadn't been wreaked on the Cartesian standpoint, Fermat before the Newtonian synthesis, and Maupertuis and Euler after it, introduced the role of extremal principles in optics and dynamics to the scientific community. Here was the reintroduction into respectable physics of something that might be called "final causes." The explanation of an optical path as one that occurs because it takes the light the least time to follow that path, and the explanation of a dynamical evolution as being that evolution that reduces the action expended over the evolution to a minimal value when compared with nearby possible evolutionary histories, introduce explanatory accounts that differ radically from several aspects of what is normally thought of as explaining by efficient causes. Events at a time are not accounted for by events at a temporally contiguous moment. Rather, a whole time series of events is accounted for in terms of a global property of that series as compared with the same global property of some related series. Nor is there the time asymmetry of explanation so familiar from ordinary causal answers to "Why?" questions. It is the whole temporal history that accounts for each part of itself, and explanations of the present are no longer being given solely in terms of reference to what happened in the immediate past.

All of this remains true even if one drops the grand metaphysical–theological claims made for such extremal explanations. Nothing in metaphysics or theology will tell us what quantity is to be extremalized (time in the case of light, action in the case of dynamics). Nor must such extremal paths be minimal in the relevant integrated quantity. They can be maxima or inflection points in which all nearby paths share the same value. Be all this as it may, the explanatory schemes certainly introduce as respectable explanatory modes ways of accounting for the facts that smack of the ancient notion of explaining by final causes. Their respectability, of course, lies, at least initially, in their success.

With the turn of the century from the 1600s to the 1700s, the debates, both scientific and methodological, soon ceased to center on Cartesianism versus Newtonianism. In the light of Newton's success with the dynamics of the cosmos, his demolition of the Cartesians' pretence to have solved the central problem of planetary motion, his success with a variety of sublunary dynamical problems treated by his methods, and the spectacular success of Newtonian dynamics in any of its forms as developed by the Bernoullis, Euler, d'Alembert, Lagrange and others and applied by them to one special problem case after another, Cartesianism soon becomes a dead letter.

But the demise of the Cartesian program, and the triumph of Newton's dynamics, does not put an end to philosophical questions and philosophical controversy. What are some of the ongoing debates that we have seen initiated in the late seventeenth and eighteenth centuries?

First of all, what should the fundamental premises of dynamics be? Newton's Second Law supplemented by a similar proposition for torques and angular accelerations? A dynamical principle of virtual work, d'Alembert's principle as it came to be called? Or, perhaps, an extremal principle such as least action, perhaps supplemented by some other rule such as the conservation of energy? It becomes clear that various fundamental principles can be adopted in the sense that, assuming one such principle, the others can be derived, but is some particular one of the principles more fundamental than the others in an explanatory sense? Or, indeed, does some "deeper" principle behind all of these lie waiting to be discovered?

What is the epistemic status and the modal status of the fundamental principle? Is it knowable a priori by pure reason, or must it be inferred by some sort of inductive reasoning from particular experiences abstracted and generalized? Are the principles necessary truths or contingent truths? We have seen Euler suggesting that $F = ma$ is necessary and Daniel Bernoulli that it is contingent. Outside the domain of the discussion of dynamics by the scientists themselves, one has such projects as Kant's rather strange attempt to show that both the Newtonian dynamical principles and Newton's gravitational law are necessary truths that are inferable a priori. This is in Kant's *Metaphysical Foundations of Natural Science*. On the other hand, one has d'Alembert suggesting that $F = ma$ is, rather, nothing but a definition of the term on the left-hand side of the equals sign.

The invocation of the notion of force in the Newtonian sense immediately leads to controversy over what such an element of the world ought to be taken to be. Is force some primitive quality of things? How do we grasp its nature in the world? What is the status of the term used to refer to it? Here we see the continuation of a long-ongoing puzzle about fundamental theories. How are we to interpret those concepts of the theory that are introduced in its explanatory apparatus in a manner that is internal to the theory, not imported into it from pre-existing language. The concepts relating to the positions of things and their variation in time seemed to most of the physicists to be unexceptionable – at least until their "absoluteness" is claimed! But how do we grasp the meaning of "force" as used in Newtonian dynamics, where it has ceased to be merely some measure of a quantity of motion? Although this was not often noticed this early in the scientific discussion, the notion of mass as employed by Newtonian dynamics, namely the intrinsic inertial resistance to change of motion, was a puzzling one as well.

It was generally assumed that concepts of position and motion, understood relationistically, were unproblematic. But, of course, the spatiotemporal concepts needed for dynamics were, according to Newton, not interpretable in a relationistic vein. The lapse of time needed to be thought of as absolute, with definite relationships of magnitude between distinct time intervals. Clocks could be better or worse indicators of absolute time, but time itself was independent of the functioning of any material clock. Position, velocity and acceleration were all absolute for Newton. They were given by the positional relation of an object to places in absolute space or the motion of the object with respect to space itself. While some material object might be at rest in space itself, and so be a marker for determining absolute accelerations, it was acceleration with respect to space itself and not with respect to any such marker that revealed itself in the inertial effects incumbent upon deviation from genuine inertial motion.

But these Newtonian notions were the subject of skepticism, even by many who were deeply committed to the Newtonian framework in general. The Cartesians, Leibniz and later Berkeley found the very notion of "space itself" philosophically objectionable. Even putting metaphysical or philosophically motivated epistemological objections to absolute space to the side, there was the deeply disturbing fact that, as Newton himself was well aware, absolute position and absolute velocity, unlike absolute acceleration, bore no observational consequences to serve as their measures. Indeed, even absolute acceleration had its epistemological problems, for, as is implicit in Newton's own corollaries to the laws, a uniform straight-line acceleration of the entire matter of the Universe would also be without any empirical consequences.

Finally there were the questions raised by the theory about the very nature of scientific explanation. What ought the model of answers to "Why?" questions be? Most of the explanatory apparatus of the theory in terms of inertial motion interfered with by an external force seemed to fit some notion of efficient cause. But even here there were questions raised about the legitimacy of such actions if they failed the test of spatiotemporal contiguity. Were "actions at a distance" legitimate explainers in any genuine fundamental theory?

But, of course, not all dynamical explanations had this "efficient causal" form. For there were also the extremal principles that accounted for dynamical evolution on what seemed to be a "final-causal" model. Were such principles ever legitimate as components of a fundamental theory? Were they revelations of metaphysical or even theological truths about the world as Maupertuis and even Euler seemed to think? Or were they, as Lagrange suggested, merely consequences of more fundamental, and far less problematic, principles formulated in the usual efficient-causal manner?

In addition there were those explanations that rested on conservation principles. Linear momentum, angular momentum and energy were all proposed as quantities whose total magnitudes never changed, although, of course, there were many cases (inelastic collisions) where energy in the form of overt motion did not seem to be conserved and other cases where its preservation required the introduction of Leibniz's "dead force" or potential energy. Where did these stand in the roster of fundamental principles of the theory? Were they primitive truths about the world or did some deeper account exist to explain their pervasiveness?

For the Cartesians, Descartes' world-picture, the "mechanical" world-view, provided an integrated metaphysical–epistemological–scientific account of nature and our knowledge of it. With the demise of this world-picture, because of its manifest failure as a scientific hypothesis, and with the rise of the Newtonian account, it was hardly a surprise that, even with its remaining conceptual difficulties, the Newtonian account would be taken by many as providing an all-encompassing scientific picture of the world, a picture that brought with it its own standards of what any good scientific explanation and any good scientific theory ought to be like.

Attempts at generalizing from the Newtonian success with the cosmic dynamics and with an ever-expanding range of sublunary dynamical problems ranged from the scientific to the metaphysical. As an example of the former there is Newton's famous speculative footnote to his *Opticks*, where it is suggested that all the familiar macroscopic properties of matter might be the result of its constitution out of microscopic parts, parts that interacted with one another by means of forces that bore some formal resemblance to the force of gravity that bound the solar system's objects together. As an example of the latter there were those suggestions to the effect that matter itself might be reconstrued as nothing but "centers of force," with force in the Newtonian sense, mysterious as that was, as the fundamental constituent of the world over and above space and time themselves. Other, quite distinct, attempts at building whole metaphysical schemes out of Newtonian dynamics range from Kant's a priorism in the *Metaphysical Foundations*, noted above, to the popular "mechanistic" world-views of nineteenth-century materialists like Feuerbach. It is something of an irony that Newton's demolition of the Cartesian "mechanistic" world-view is taken as the basis of a world-view that is itself often labeled "mechanical."

CHAPTER 13

Conservation principles

13.1 DISCOVERY, CONTROVERSY AND CONSOLIDATION

As is the case with so many of the fundamental concepts of dynamics, Galileo provided the first seminal ideas. In his discussion of falling bodies constrained to inclined planes, he offered an ingenious argument to the effect that the speed obtained at the end of the fall will be dependent only upon the height of the fall, and not on the slope of the incline, at least insofar as friction and air resistance can be ignored.

The argument rests upon intuitive agreement that an object suspended by a string from a pivot point and allowed to fall from a certain height will return to the same height even if the string encounters a nail around which it must pivot at some point in the descent. Two deep notions are encountered here. One is that height in a gravitational field is the unique parameter needed in order to capture the potential for a fall from that height to be able to generate a specific quantity of motion. Here we have the beginnings of the notion of potential energy. The other idea is that the motion generated by a change in height will be independent of the path by which that loss of height is obtained. Here is the beginning of that essential aspect of potential energy that the change in potential energy on going from one point to another, and hence the quantity of motion generated or absorbed, is path-independent. Galileo is clearly aware of the essential difference between the energy "stored" by a gain of height and the energy "lost" owing to, say, friction of the inclined plane or air resistance. In the former case the quantity of motion that disappears can be made to reappear merely by restoring the object to its original height. In the latter it cannot. In the former case the motion temporarily lost is path-independent, depending only on the height gained. In the latter the motion that vanishes is path-dependent.

But the first intense attention paid to conservation laws came from that other great problem of early dynamics, the problem of collisions. Here the seminal work is that of Descartes, replete with confusion as it might be. Simple observation of collisions suggests that the motion lost by one colliding object is picked up by the other. Clearly the size of the colliding object is playing a role as well, although originally, of course, the notions of weight and of inertial mass had not clearly been separated. In Descartes' case there is also the confused desire to equate mass with some purely geometric quantity such as the volume of the object in question. Descartes' law of conservation of motion is, as we have noted, badly flawed. For he takes the sum of the quantity of motion to be conserved, where for each object its quantity of motion is given by its mass times its undirected speed. This leads Descartes into his unfortunate collection of laws of collision, all but one of which are incorrect and which form an inconsistent whole.

It is Descartes who combines the principle of conservation of motion with his overall dynamical view of the world as a plenum of moving matter with motions being generated by contact transmission. The entire realm of dynamics is then subsumed into the one principle that matter is initially instilled with a quantity of motion, which then remains unchanged in perpetuity. Here conservation of quantity of motion, together with the inertial principle, is made the central law of all dynamics.

J. Wallis in his treatment of inelastic collisions realized, at least for the one-dimensional case, that it is *directed* motion that must be taken into account when quantity of motion is taken to be conserved. His work was followed up by C. Wren.

But the real forward step was that of Huyghens in his brilliant treatment of elastic collisions. We saw how he introduced the device of solving a problem by transforming from one inertial reference frame to another, thereby allowing asymmetric cases to be treated by taking the symmetric case as solved by its symmetry. In his treatment of the elastic collision of two bodies, this essentially takes as a law the conservation of momentum, with momentum now taken as the product of mass and directed velocity. But he adds to this the conservation of the height-obtaining ability of the motions, thereby adding a conservation-of-energy principle to that of conservation of momentum. Taken together, these two principles allow the complete solution of this particular dynamical problem.

Leibniz is insistent that the quantity of motion is measured by mass times the square of velocity. Here it is once again Galileo's principle about something falling from a height being able to re-attain that height which is the inspiration, for the quantity of motion associated with fall from a given height is the square of the terminal speed obtained, not the speed itself. John Bernoulli, following Leibniz, propounded the principle of conservation of "live force," measured as kinetic energy. Daniel Bernoulli showed how one could make brilliant use of the principle by developing his famous formula relating the velocity of a fluid to its pressure solely by ingenious application of the principle of conservation of energy.

The early eighteenth century saw a tangled debate over the issue of "*vis viva*." Was the correct measure of quantity of motion mass times speed, or mass times the directed speed or velocity, or was it mass times the square of speed, the kinetic energy? The debate, with all sides buttressing their arguments with a-priori postulates of the most egregious "metaphysical" kind, suffered badly from the misuse of particular examples and the particular confusions endemic upon reliance upon Descartes' fallacious use of scalar speed instead of directed velocity when calculating the quantity of motion.

With the fall of the Cartesian picture it became very dubious indeed that conservation of quantity of motion, no matter how calculated, could be the centerpiece of an adequate dynamical theory. Newton placed force, the force derived from statics by Galileo, at the center of his theory. Without some constitutive rule of force, such as the law of universal gravitation, no full dynamics could follow. On top of that, there was all the evidence of the dissipation of energy in inelastic collisions or where frictional forces were involved, and of the generation of new motion by the conversion of potential to kinetic energy. This cast grave doubt on the universal applicability of the conservation of motion in the sense of the conservation of kinetic energy.

Gradually it became understood that in many cases, that of elastic collisions for example, or where the forces between the objects could be generated as the gradients of a scalar field, that is to say where, as in the gravitational case, a potential existed for the forces, both the conservation of momentum (using directed velocities) and that of energy were viable principles if potential energy was taken into account.

Finally, as we have noted, in his work on linked rods, rotating bodies and continua, Euler realized that an additional conservation principle was fundamental. As we have seen, Newton had already utilized a special case of the conservation of angular momentum in his derivation of Kepler's area law from the mere fact that the force governing the planets was directed toward a fixed center. At the hands of Euler this became generalized into the general law of the conservation of angular momentum.

An important step forward was d'Alembert's demonstration that in particular cases the law of conservation of energy could be derived from his fundamental dynamical postulate, the dynamical generalization of the principle of virtual work. In the hands of Lagrange this idea, namely that the conservation principles were all mere derivable consequences of the general dynamical law, became a central theme.

We noted how Lagrange prefaced his work on dynamics in his *Analytical Mechanics* by an historical survey. There he stated four principles that had previously been suggested to be fundamental: (1) the conservation of energy attributed to Huyghens, Leibniz and John Bernoulli; (2) the principle that the state of rest or motion of the center of gravity of a system of interacting particles remains unchanged by the forces of interaction attributed to Newton, with a generalization by d'Alembert to include systems of particles all subject to the same external accelerating force, either constant and acting along parallel lines or directed to a point and distance-dependent; (3) the principle of the conservation of angular momentum for the motion of several bodies about a fixed center, a principle Lagrange took to have been discovered simultaneously by Euler, Daniel Bernoulli and Patrick d'Arcy; and (4) finally, the principle of least action discovered by Maupertuis and made rigorous by Euler.

After presenting his basic law of dynamics, the fully general version of the principle of virtual work for dynamics, Lagrange proceeds to demonstrate the derivability of each of these four principles from his fundamental law. He carries out this work prior to going on and showing how his method can be made suitable for application to constrained systems when the constraints are holonomic and arriving, when the forces are generable from a potential, at his version of what later became called Lagrange's equation.

We have noted how his derivation of the least-action principle serves for Lagrange, and for many who follow, as a "demystification" of this principle that seemed to introduce the dubious notion of final causation into physics. For each of the other principles Lagrange's aim seems to be simply to show that the various "special" principles posited by others all follow, in the appropriate circumstances, from his single unifying general dynamical principle.

For the motion of the center of gravity, one considers a set of particles interacting with one another. Since the forces of interaction depend only on the relative positions of the objects with respect to one another, the specific position of any one particle, to which the others are then referred, will drop out of the equations. From this it will follow that the motion of the center of gravity of the system remains uniform. This amounts to showing that, for systems of the kind considered, linear momentum will be conserved. A similar argument using the relative angular positions of the components of a system with respect to one another, and the independence of the forces on the chosen reference axes, leads to the "area law," that is to say, to the principle of the conservation of angular momentum.

Things are a little more complicated with respect to the issue of the conservation of "*forces vives*." Suppose the forces that appear in the equations are all derivable as gradients of a single scalar function of position. That is to say, suppose that all the forces can be derived from what came to be called a potential energy function. Then the sum of the potential energy, derivable from the positions of the components of the system, and the kinetic energy, derivable from the squares of their velocities, can be shown to be invariant over time from Lagrange's basic equation. But what is the case for other systems? It will hold, he argues, for "inelastic fluids, so long as they form a continuous mass and there is no point of impact between their parts." In the case of collisions it will hold for the impact of "elastic bodies," for in those cases it can be argued that the forces that generate the changes of motion by compression and then re-expansion of the colliding bodies restore to the overt motion any energy that had temporarily been removed from it during the compression phase. But, of course, this will not be the case for bodies that suffer permanent changes of shape due to the collision, namely inelastic bodies.

The status of the conservation of energy as a general principle became a major area of scientific inquiry in the century following Lagrange. Here one needed to move outside considerations peculiar to dynamics properly so called. The development of thermodynamics, and later the developments in the atomic constitution of matter, in the statistical-mechanical theory underlying thermodynamics, and in the nature of substances of a non-atomic nature such as electromagnetic fields, all played crucial roles. What was needed was the insight into the fact that, alongside the kinetic energy of overt motion and the potential energies of such things as gravitational and electromagnetic forces and the forces engendered by stresses on material such as those of tension and compression, one needed to take account of the internal energy of substances that could be transmitted from one substance to another as heat flow and of the energies stored in the field magnitudes themselves.

Some of this energy did turn out ultimately to be a form of kinetic energy of motion of the microscopic components of macroscopic matter, "the quantity of motion that we call heat," as Clausius characterized heat in a famous phrase. Other energy could appear not as simple energy of motion or as simple potential energy but as a new kind of quasi-kinetic energy of the fields themselves (for example, the energy of a beam of electromagnetic radiation). What was crucial was the realization, starting with the careful experimentation of Joule and others, that a true conservation-of-energy principle that was exceptionless could be devised so long as one broadened one's horizon beyond that encompassed by standard dynamics in its narrow sense.

But these important developments in physics will be, for the most part, beyond our scope in this book. We will, however, return to related issues when we discuss some of the controversy between "atomists" and "energeticists" in the latter part of the nineteenth century, insofar as that controversy impinged on the development and interpretation of dynamics.

13.2 SYMMETRY AND CONSERVATION

In the nineteenth century much of the discussion about the importance of conservation rules centered on the issue of the universality of the conservation of energy that we have just noted. This discussion tied in with the issue of the discovery of the fundamental role of time-asymmetric entropy increase in thermodynamics. A model of the world in which energy was always conserved, but in which the total energy of the Universe was ever more uniformly distributed over all of its matter, became popular.

Speculation was rife about the ultimate end of the Universe with uniform temperature everywhere, the so-called "heat death" of the Universe.

Meanwhile, within dynamics proper, the laws of conservation of momentum and of angular momentum played crucial roles in the application of dynamics to more complex systems, especially to fluids and to more complicated continuous materials.

In the case of fluids the conservation rules, along with the conservation of the fluid matter itself, provided the crucial constraint equations needed in order to provide the full partial differential equations of fluid dynamics. In more complicated continuous media, the conservation rules played a role in constraining the rules for the kinds of forces that the material could suffer. Here one sees a process that generalizes matters going back to Newton. We saw that Newton's Third Law, the law that every action has a reaction that is equal in magnitude and opposite in direction, could be considered an early version of the principle of conservation of momentum. In the special case of forces that particles exerted upon one another and that were always directed along the line connecting the particles, Newton also realized that the principle of conservation of angular momentum followed from the law.

In the more general perspective, matters went the other way. One assumed the truth of the principles of conservation of momentum and conservation of angular momentum. One then took these as constraints upon the forces the components of the continuous media could exert upon one another. The result was that these forces would have to obey various symmetry conditions. If one component exerted a direct force on another, this had to be matched by an equal opposite force. In the case of the general notion of stress, things became moderately complicated. The changing shape of an object would generate various internal forces that the infinitesimal components of the object exerted upon one another. This led to the notion of the stress tensor. Changing the dimension of the object along one axis could generate forces in any direction. These forces could be decomposed into their components along a set of orthogonal axes, as could the original dimensional change. The stress tensor would relate the components of the forces so generated to the components of the dimensional changes. The imposition of conservation of angular momentum forces this tensor to be symmetric, thereby greatly constraining and simplifying the production of the general partial differential equations and their solutions. Even when the forces were taken to be discontinuous, as in the case, for example, of the dynamics of shock waves, the imposition of the conservation rules was essential to characterizing the dynamics.

Deeper insight into the place the conservation rules really played at the foundations of dynamics came with reflection upon the kinds of arguments Lagrange had used in deriving the principles from his fundamental dynamical law. The conservation of momentum follows when the forces in a system depend only upon the relative separations of the components and not upon the place "in space" of the system as a whole. The conservation of angular momentum holds when the forces involved depend only upon the relative orientations of the components of the system, and not upon the orientation "in space" of the system as a whole. And the conservation of energy holds when the forces involved do not depend upon the place "in time" in which the process is occurring.

Or, to put it more explicitly in physical terms: Conservation of momentum will hold when, if a system evolves in a certain manner, then a similarly constituted system that is different from the original only in that all the components of the system have been identically translated to some new place relative to the older system will behave in the same manner. Conservation of angular momentum will hold when, if a system behaves in a certain manner, then a similarly constituted system that is different from the original only in that the system as a whole is a rotated version of the original system will behave in the same manner. And conservation of energy will hold when, if a system behaves in a certain manner, then a similarly constituted system whose initial state is identical to that of the original system, except for being earlier or later than the initial state of the original system in time, will evolve with its states being identical to those of the original system except that they are also shifted in time by the same amount from the evolving states of the original system. In Lagrange this last condition is hidden in the fact that the potential from which the forces are derived is itself not explicitly dependent upon the time.

Mathematically all of this may be dealt with elegantly by a theorem due to Emmy Noether. Let a dynamical system be described by a smooth (i.e. infinitely differentiable) Lagrangian function. In the case of "natural" systems this will just be the kinetic energy of the system minus its potential energy. Suppose there is mapping such that the Lagrangian function remains the same function of the coordinates under the mapping. Then the Lagrangian system is said to "admit the mapping." Suppose there is a one-parameter group of diffeomorphisms (i.e. such smooth mappings) admitted by the Lagrangian system. For each such group of mappings there is a first integral of the dynamical system. That is to say, there is a function of the generalized coordinates and velocities that is constant in time.

Consider some examples. Suppose the system is invariant under translations along some axis in space. The one-parameter group of diffeomorphisms admitted by the system is just the translations along the axis parameterized by the distance the system is translated. In this case the linear momentum along that axis will be a constant of motion of the dynamical system. If the system is invariant with respect to rotations around some axis, parameterized by the angle of rotation, then the angular momentum around that axis will be a constant of the motion. And, if the system is invariant under a shift in time, with the duration of that shift as parameter, then the total energy of the system will be conserved.

All of the standard conservation principles, then, appear as consequences of this one unifying theorem.

If a system is embedded in a potential thought of as external to the system, and if that potential fails to have the relevant symmetry, then the system will not, of course, have the associated conserved quantity. But suppose we are considering only forces generated from potentials dependent on the structure of the system itself. A group of particles, for example, may be thought of as moving under forces solely due to the gravitational interactions among the particles themselves. Then where the system is in space, what its orientation is in space, and when in time the system is evolving will be irrelevant to its dynamical evolution. So its momentum, angular momentum and energy will be conserved.

This is assuming, though, that "absolute" position in space, "absolute" orientation and "absolute" position in time are irrelevant to the dynamics. In other words, what the mathematics is telling us is that the origin of the conservation principles is to be found in the homogeneity of space (space being "the same" at every point), the isotropy of space (space being "the same" in every direction) and the homogeneity of time (time being "the same" in every interval).

The conservation rules provide us with a very nice example of how the status of principles can change within a theory as that theory evolves over time. Originally these rules arose from generalizing on experimental revelations in particular cases. Falling objects were able to regain exactly the height from which they fell. In collisions the loss of motion of one object seemed to be compensated for by the gain in motion of the other. Soon it was proposed that some conservation law was the central law governing all of dynamics, as in the Cartesian account. For some time, of course, controversy raged over the correct measure of the conserved "*vis viva*" or force of motion, with Descartes' scalar speed, vectorial velocity and velocity squared (each multiplied by the "amount" of moving matter eventually encapsulated in the notion of inertial mass, of course) all contending.

Later the Lagrangian attitude became the dominant theme: The fundamental law governing motion was that of the dynamically generalized principle of virtual work. From this principle, in special cases, conservation principles, as well as extremal principles, could be derived as mere consequences of the underlying fundamental law. The conservation rules played a fundamental role in solving dynamical problems in that they provided constraints upon the forces involved (at least in the case of the conservation of momentum and angular momentum), thereby greatly restricting the possibilities for the constitutive equations that specified the exact forces involved in the dynamics.

With the realization of the role played by symmetry considerations in determining the contexts in which the laws could be applied, however, there was finally achieved the deeper understanding that the conservation rules each represented, as instances of a single general theorem of Noether, some posited symmetry of space or time themselves.

This new understanding of the, once more fundamental, place of the conservation laws in the theory was important when physics moved beyond Newtonian dynamics to the newer relativistic theories and to quantum mechanics. In the formulation of special relativity, for example, Einstein makes the fundamental assumption that the equations transforming descriptions of motion from one coordinate frame to another moving with respect to it, the Lorentz transformations, are linear. These assumptions amount to assuming flatness of the underlying, Minkowski, spacetime. Flatness implies translational symmetry and rotational isotropy, giving rise to new conservation rules for special relativity, which, while distinct from those of Newtonian dynamics, play a similar foundational role in the new theory.

Symmetry principles, and their correlated conservation rules, also play a fundamental role in the foundations of quantum mechanics. Here the familiar spatio-temporal symmetries may be posited with their corresponding energy-, momentum- and angular-momentum-conservation principles. These are then often supplemented by additional posited symmetries (some of which, unlike the spacetime symmetries, are discrete rather than continuous) such as symmetry under reflection (parity conservation), under exchange of particles for their anti-particles (charge conjugation) and under time reversal. In profound generalizations of the Noether results one can, for example, relate the charge-conservation principle of electrodynamics to symmetry under a new transformation, namely the unitary symmetry of the electromagnetic field.

Interestingly, the discovery of general relativity, with its postulation that the spacetime of the world may be curved, and variably curved at that, places new limitations on the applicability of the older spatio-temporal conservation principles. In general relativity there is a kind of "local" conservation principle that always holds: the covariant divergence of the stress-energy tensor is zero. Since the equations of the theory relate the local condition of the spacetime to the stress-energy tensor, this imposes a deep constraint on possible spacetimes. But global conservation principles depend upon the symmetry of the spacetime, and in general-relativistic worlds such symmetries need not exist. Thus one will have conservation rules for momentum and angular momentum only in worlds that are, at least in the asymptotic limit far from matter, translationally or rotationally symmetric. The conservation of energy is also a very problematic notion in general relativity, since energy, in a sense, can feed into and out of the curved spacetime (the gravitational field) itself. The result is a limited possibility of imposing an energy-conservation principle, but only with the inability to localize the energy of the spacetime curvature itself, to say where in the spacetime the energy is.

The changing status of the conservation principles gives us a nice example of how a fundamental principle can radically change its status within the foundational theory over time. Initially there was the hope that the conservation principles could, by themselves, ground all of dynamical theory. This conception died with Newton's demonstration of the necessity of specific force laws in solving dynamical problems. The conservation principles seemed even less fundamental when Lagrange derived them from his underlying principle that was based on the dynamical extension of the virtual-work concept. With Noether's results the fundamental basis of conservation principles in spacetime symmetries was revealed. With that understanding, conservation principles could be seen to be sufficiently general that they might be applied even outside the original dynamical theory in which they arose. Consider, for example, the crucial role played by the symmetry considerations in Einstein's discovery of a new dynamics appropriate to the Minkowski spacetime needed to do justice to optical phenomena. Finally, the limits of the global conservation rules appear in general relativity, but even there local conservation remains in force.

SUGGESTED READING

Chapter IV, Section 3 and Chapter VII in Part II of Dugas (1988) describe the early discovery of conservation laws; Chapter V describes how conservation laws were employed by Huyghens, Wallis and Wren. For accessible treatments of how Noether's theorem ties symmetry to conservation see Hanca *et al.* (2004) and Neuenschwander (2010).

CHAPTER 14

Hamilton's equations

14.1 THE HAMILTONIAN FORMALISM

The nineteenth century saw several new proposals for fundamental principles to be placed at the foundations of dynamics. In 1829 Gauss offered the "Principle of Least Constraint." Let a system of particles be connected together by constraints. Suppose each particle n starts at A_n and after a time interval ends up at B_n. Let C_n be the position it would have ended up at at the end of the time interval had the constraints not been imposed. Then, for the actual motion undergone, the sum of the mass of each particle times the square of the distance from B_n to C_n will be minimal over the class of all motions compatible with the constraints binding the particles together.

In 1894 Hertz offered the "Principle of Least Curvature." Generalizations of the ordinary Euclidean notions of distance along a path and curvature of a path are constructed. It is then shown that, using these definitions, a system of particles that is isolated, that is not subject to some external force, will evolve in such a way that the motions of the individual particles will generate a path in a multi-dimensional space for the point representing the system as a whole that has, at each point, minimal curvature in the sense defined by Hertz. This principle provides a useful opening for applying methods of differential geometry in dynamics, for in a curved space the paths of least curvature, the geodesics, are objects of intensive study in differential geometry. Hertz's principle then tells us that, with the proper definitions of distance and curvature for our space, one can view the dynamical trajectory of a system as a geodesic in the representing space.

Far more important than these principles, though, is the discovery by Hamilton of a new way of representing the dynamical equations of motion. Let L be the Lagrangian function, that function which appears in Lagrange's equation, which is derivable either from a principle of virtual work or from a principle of least action. In the case of a conservative system it is just the kinetic energy of the system minus its potential energy. This function is a function of the generalized position coordinates, the generalized velocities and the time. Conjugate to each generalized coordinate, a generalized momentum is defined as just the partial derivative of the Lagrangian function with respect to the generalized velocity associated with the generalized coordinate in question. If the coordinate is just a simple linear distance, the associated momentum is the usual mass times velocity in that direction. If the generalized coordinate is an angle around some axis, the generalized momentum is just the angular momentum around that axis. In more abstruse cases generalized momentum won't look like any of the familiar "quantities of motion."

Next define a new function of the generalized coordinates, the generalized momenta and the time, the Hamiltonian function H. H is just the sum over all of the coordinates of the generalized velocity for a coordinate times the generalized momentum for that coordinate, minus the Lagrangian function. In the neatest cases, where the system is conservative with a potential not depending on velocities, and the constraints are independent of the time, H turns out to be just the sum of the kinetic energy and the potential energy of the system. That is, the Hamiltonian is just the total energy of the system. All of this can be generalized to deal with cases where the potential function is also velocity-dependent.

The Hamiltonian function is related to the Lagrangian function in a profound way. The Hamiltonian function is the Legendre transformation of the Lagrangian function. A Legendre transformation takes a function f of variables $x_1, \ldots, y_1, \ldots$ and changes it into a function $g = f - \Sigma_j x_j p_j$, where p_j is just the partial derivative of f with respect to x_j. g is then a function of the y variables and the derivatives of f with respect to the original x variables. Such transformations are used with abandon in thermodynamics to convert descriptions of systems from dependence on one set of thermodynamic variables to those invoking some other set. If g is the Legendre transformation of f, then f is the Legendre transformation of g.

Once one has the Hamiltonian function, the fundamental equations of dynamics can be written in a new way. Instead of a single second-degree differential equation of each generalized coordinate and velocity, one now has a pair of first-degree equations for each generalized coordinate and its conjugate generalized momentum (and one more equation if the Lagrangian – and hence the Hamiltonian – had explicit time dependence in it). The time rate of change of a generalized coordinate is just the partial derivative of the Hamiltonian function with respect to the conjugate momentum; and the time rate of change of a generalized momentum is just *minus* the partial derivative of the Hamiltonian function with respect to the associated generalized coordinate.

As in all the other cases where a new formalization of the fundamental laws becomes available, the Hamiltonian formulation provides a new resource for conveniently characterizing and solving dynamical problems. One set of cases where use of the Hamiltonian formalism is effective is those in which the Hamiltonian function does not depend upon one of the generalized coordinates. The coordinate is said to be "cyclic" in this case. In the Lagrangian formalism, even if the Langrangian function is independent of some generalized coordinate, one may still have to deal with the explicit time dependence of the generalized velocity associated with that coordinate. But the second of Hamilton's equations shows in a trivial fashion that, if a generalized coordinate does not appear in the Hamiltonian function, the corresponding generalized momentum is just a constant of the motion.

The Hamiltonian formalism also serves to bring to the surface the important fact that the function that gave the total energy of a system as a function of its generalized coordinates and momenta was also the function that characterized the "infinitesimal" way in which these coordinates and momenta changed in time. Given a description of a system at one time, then, it was the way in which the total energy depended upon the fundamental variables that determined how the system would evolve with time. The Hamiltonian is, then, the "generator" of time translation for the system. It was soon realized that the linear momentum function played an analogous role as generator of spatial translations and the angular momentum function as generator of rotations for the system. This fundamental role for these quantities, which is plainly connected with the association of their conservation with symmetry in the respective transformation of a system, plays a crucial role when the structure of classical dynamics is used to build the framework for quantum mechanics.

14.2 PHASE SPACE

The manner in which the new Hamiltonian formalism had its greatest impact on dynamics, though, is in the introduction of the notion of generalized momentum. This new concept does not, of course, introduce into the world some new feature that is not implicitly taken account of in the previous formulations of dynamics. "In principle" everything you can find out about the world using the notions of generalized position and momentum could be described using the ordinary notions of space and time, of position and velocity and their changes, or the generalized notions of position and velocity that are used in the Lagrangian formulation. But the new concept provides a framework for further research in which much that was hidden, and might have remained hidden, becomes exposed for fruitful further theoretical understanding.

Although there is a clear sense in which position and time are the fundamental kinematic quantities, with such things as velocity, acceleration and momentum – ordinary or generalized – being derivative concepts, for many purposes it is extraordinarily useful to think of dynamical systems as characterized by their generalized coordinates and their generalized momenta, with these as the "independent" variables describing the system. This way of dealing with systems, looking at how these two sets of generalized variables evolve over time, gives insights into dynamics, and into the relationship of dynamics to other aspects of physical theory, that are obtainable in no other way.

A graphic way of providing such descriptions is through the use of phase space. Here one uses the mathematical artifice of a multi-dimensional space, each point of which specifies a possible set of all the values of the generalized coordinates and the generalized momenta describing a system. As the system evolves, and the values of the generalized variables change, a point describing the system follows a one-dimensional trajectory through the multi-dimensional space. The description of a system, or a collection of systems, by such a phase-space trajectory, or set of such trajectories, becomes the bedrock on which much of contemporary dynamics is based.

We shall make much use of the phase-space description of the evolution of systems later when we take up the development of dynamics from the attack on the problems of celestial mechanics, through the discovery of qualitative dynamics and on to the invention of chaos theory. For now, however, I will focus on just one feature of the phase-space picture of dynamics, and how focussing attention on that feature has its important consequences for understanding in branches of physics unimagined by Hamilton when he first constructed this new formalization for dynamical theory.

One of the many discoveries in dynamics of H. Poincaré was that of integral invariants. A collection of systems, each isolated from its environment, is represented at a specific time by a collection of points in the phase space, the points characterizing the states of the system at the time. After an interval of time a new collection of points will represent the states of the systems at the later time. Suppose we measure the volume of the collection of points at the first moment of time in a very natural way, using the "extent" of the region in each of its dimensions corresponding to the position and momentum coordinates of the phase space. Then Poincaré's work shows that, along with other definable quantities, this "phase-space volume" for the set of points at the first moment of time will be the same as the volume of the set of points representing the states of the systems at the later time. This invariance of phase volume follows from the fact that the system is represented in terms of generalized coordinates and momenta that obey the Hamiltonian dynamical equations.

This invariance of phase volume over time becomes a central pillar of the theory of statistical mechanics. Thermodynamics describes macroscopic systems in terms of a small number of parameters such as volume, pressure, temperature and entropy. Systems have their behavior governed by a few general laws, the First and Second Laws of thermodynamics that stipulate the conservation of energy and the time-asymmetric behavior of entropy changes in particular, and by specific "constitutive equations," such as the ideal-gas law, that govern the inter-relations of the parameters specific to a particular type of system. Kinetic theory tells us that the macroscopic systems are made up of innumerable microscopic components, such as the molecules of a gas, and that the behavior of these components is governed by the fundamental dynamical laws. In its early versions these dynamical laws were, of course, just those of the Newtonian dynamical theory.

But how is the macroscopic characterization of the system and its law-like structure to be related to the dynamics governing the behavior of the micro-components of the system? In the course of a long series of innovative developments at the hands of Maxwell and Boltzmann, and in response to a series of incisive critical doubts about the original theory proposed by a number of distinguished theoreticians, it became clear that the theory needed to be formulated in terms of a consideration of "probabilities" obtaining for any collection of the microscopic states of the system compatible with the macroscopic constraints imposed upon it. It was in assigning such probabilities and in determining how these probabilities should be considered to evolve with time that the fundamental connections to the macroscopic regularities of thermodynamics were to be found.

Clarifying this from a philosophical perspective amounts to a major investigative program in its own right, one we certainly cannot pursue here. But a few brief remarks will highlight the theme which is important for our present project. What seemed, initially, to be merely an elegant reformulation of the fundamental laws of dynamical theory in a novel mathematical way resulted in the exposure of conceptual resources that had profound ramifications for the ability to apply the dynamical theory in ways totally unexpected either within pre-Hamiltonian dynamics or even by Hamilton himself when he discovered this novel set of dynamical equations. The resolution of the usual second-order dynamical equations into a pair of first-order equations resulted in the introduction of the generalized momenta paired with the generalized coordinates, instead of the older reliance on generalized coordinates and generalized velocities. The description of the system in terms of generalized coordinates and momenta resulted in the idea of phase space as the proper arena in which to locate the descriptions of the momentary states of systems (as points in phase space) and the descriptions of the ways in which systems evolved over time (as one-dimensional trajectories through phase space). But the introduction of the phase-space description led to the discovery of the integral invariants, including phase-space volume, by Poincaré, and this in turn provided the resource needed to make inroads into the program of connecting probabilities over microscopic states and their dynamics with the thermodynamically described macroscopic behavior of systems.

In thermodynamics a special state of systems is distinguished, the equilibrium state. Systems not in this state have their parameters change. Systems in this state remain unchanged over time. One branch of statistical mechanics, equilibrium theory, identifies the appropriate quantity to associate with macroscopic equilibrium as a probability distribution over the microscopic states of systems that, as the individual systems change with time in the manner required by the dynamical laws, itself remains invariant in time. The picturesque description is of a collection or "ensemble" of systems all subject to the same macroscopic constraints. The individual systems may have any of the microscopic conditions compatible with these constraints. One thinks of the systems as occurring in the collection with such frequencies that the number with their microscopic conditions within a given collection of such conditions is specified by the probability distribution used to characterize equilibrium. Since equilibrium is itself an unchanging condition in time, one wishes to associate with it a probability distribution that is such that as many systems will enter some specified collection of conditions over time as leave that collection, with the overall distribution of systems remaining unchanged.

Here the invariant phase volume discovered by Poincaré becomes essential. For it is probability *uniformly* distributed over the phase-space region that contains all of the possible micro-states compatible with the constraints that then provides the invariant probability needed to describe equilibrium. What does "uniformly" mean here? That is a notion derived from the "natural" way of measuring volumes in phase space, the so-called Lebesgue measure. Actually, for the systems with which the theory is concerned, say those with a fixed constant energy, a subtle modification of the most natural measure is needed. But, basically, the point is that, by using the results about invariants in phase space under evolutions described by the Hamiltonian equations of dynamics, one is able to construct the desired time-invariant probability over microscopic conditions that one wants in statistical mechanics to associate with the equilibrium macroscopic state of a system in thermodynamics.

Many philosophically interesting puzzles raise questions about the nature and legitimacy of this entire branch of statistical mechanics. The very kind of "explanation" of the phenomena being offered is deeply problematic. Furthermore, there remain additional questions about the uniqueness of the stationary probability measure derived as just described. How does one know that the probability obtained as that uniform over the phase-space region in the relevant sense is the only one which is unchanging in time? The answer to that question is often taken to lie in what is called ergodic theory. Here profound considerations of the way in which the trajectories mapped out in phase space as systems evolve under their dynamic constraints behave in a sufficiently "disorderly" manner become important. One needs to show, according to this account, that trajectories from "almost all" initial states will "wander randomly" over the entire available phase space. We will discuss the way in which this issue becomes another previously unsuspected entire new branch of dynamics when we take up issues of stability and instability of dynamical systems in a later chapter.

The other branch of statistical mechanics is that which tries to do justice to non-equilibrium phenomena. Here, once again, the theory often tries to represent the macroscopic phenomena, in this case the way in which the macroscopic thermodynamic parameters change as a system prepared in a non-equilibrium state approaches equilibrium, by looking at the behavior of a probability function defined over the points of the phase space of the system. One approach first takes probability to be distributed uniformly over all the points that can represent the system when it is subjected to some initial constraints. Then those constraints are changed. Imagine, for example, a gas confined to the left-hand side of a box by a partition, which is then removed. The system is now in a non-equilibrium state. Now a new, larger, set of points in phase space represents the new set of phase points that can correspond to the possible micro-states of the system under the new, changed, constraint condition. Can one show, using the nature of the microscopic constituents and how the macroscopic object is made up out of them, and using the dynamical laws governing the microscopic elements, that the initial probability distribution will "flow" in the phase space to become the probability distribution for the system that corresponds to the equilibrium it will obtain given the changed constraints? Actually one can show, using the invariance of phase volume under the dynamics, that it won't. Much of non-equilibrium statistical mechanics consists of trying to find some weaker sense in which the probabilities will change under the dynamics so as to represent the approach to equilibrium, or to characterize within the probabilistic theory that approach in some other subtle manner.

Here we cannot pursue this rich, and philosophically rewarding, area of foundational physics. But what is relevant to our purposes here is the light all of this throws on the place Hamiltonian dynamics plays in the overall development of dynamics. Here we see significant changes in the motivation and the consequences of the development of the foundations of the theory from those that characterized the work of the late seventeenth and eighteenth centuries.

In the late seventeenth and eighteenth centuries two goals motivated the search for fundamental principles of dynamics. First, there was the desire to solve certain particularly difficult and crucial problems. How could the Keplerian orbits of the planets be accounted for? How could the dynamics of the compound pendulum be captured? How could one deal with constrained systems without figuring out all of the forces of constraint? How could the motion of rigid bodies be characterized? How could the dynamics of fluids be encompassed? Second, there was the desire to find truly fully general first principles from which all of the detailed laws of dynamical systems could be derived.

It was in pursuit of these joint goals that the various foundational principles were discovered. The fully general principle of forces and torques generating linear and angular accelerations of Newton and Euler, the dynamically generalized principle of virtual work of James Bernoulli, d'Alembert and Lagrange and the extremal principles of Maupertuis, Euler and Lagrange were all found during this incredibly fruitful century-long period of discovery.

In the nineteenth century the program of applying the dynamical laws to ever more complex and difficult systems continued. There is, for example, the extended program of dealing with continuous media more complex than simple fluids and the program for dealing with complicated heavenly motions in which multiple attracting bodies are involved known as celestial mechanics. But it is taken for granted that the fundamental general principles of mechanics are now known. Indeed, there is clear awareness that one could pick any one of the three basic fundamental principles as the foundational law and derive the other principles from it.

In deriving his equations of motion, Hamilton sought only to provide a mathematically interesting alternative to the existing foundational principles. There was no thought that these principles lacked something in generality that the new formalism could provide, and Hamilton's new formalism proved of little value in facilitating the applicability of the fundamental principles to particular problems. Unlike, say, the Lagrange equations for dealing with constrained systems, characterizing the dynamics of systems by Hamilton's paired first-order equations usually led, quickly, to the same equations as those which needed solving when the problem was treated by the familiar methods, say by the use of forces and torques, masses and moments of inertia or by Lagrange's virtual-work methods.

The true value of the Hamiltonian formalism, rather, lay in its ability to open up to view structural elements already implicit in the theory but hidden from sight in its earlier formalisms. Typical of this was the revelation of the important role in revealing structure played by the notion of a generalized momentum conjugate to a generalized position coordinate. We have seen how this led to the introduction of the phase-space picture of dynamical evolution and, ultimately, to an invaluable component of statistical mechanics. As we pursue the other advances of dynamics in the nineteenth and twentieth centuries, we shall see how Hamilton's reformulation of the theory led to still further reformulations that played the same crucial role. Although each of these new formulations of the theory had some value in increasing the problem-solving capacity of the theory, their real contribution to the advancement of physics was to provide ever deeper understandings of the hidden structures of the theory and to provide conceptual devices that allowed the theory to be extended to novel and previously unexpected domains.

SUGGESTED READING

For Gauss's Principle of Least Constraint and Hertz's Principle of Least Curvature see Chapter IV, Section 8 of Lanczos (1970). For an outline of the history of Hamilton's equations and for an introduction to phase space and to Poincaré's integral invariants see Chapters VI and VII of Lanczos (1970). For an outline of the development of the role of phase space in statistical mechanics see Chapter 2, Sections II–IV of Sklar (1993). The role of phase space in equilibrium theory can be found in Chapter 5 of the same book.

CHAPTER 15

Canonical transformations, optical analogies and algebraic structures

In this chapter we will explore three more stages in the nineteenth-century development of dynamics. One program was not itself directed at finding new foundational posits for the theory. Its initial purpose, rather, was to supply a method to facilitate the solving of dynamical problems, especially when they were framed in the mode of the Hamiltonian dynamical equations. We need to pursue it a bit, however, since it provides some of the resources needed to understand the second program treated in this chapter.

This second program is Hamilton–Jacobi theory. Here, starting from Hamilton's work in optics, rather than in dynamics, the result was the development of new possible foundational equations for dynamics to supplement those already known. Just as in the case of the development of the Hamilton dynamical equations, there was no claim here that the results went beyond the existing foundational posits in any fundamental way. It was universally accepted that the existing foundational methods were correct and complete as they stood. Rather, a new "reformulation" of the existing foundations was what was on offer.

As was the case with the Hamilton dynamical equation, the new foundational equation, the Hamilton–Jacobi equation, does provide a useful resource in problem solving. There are some cases where problems approached by employing the new method receive more facile solutions than could be obtained using the older foundational equations. But, again as in the case of the Hamilton equations, the primary value of the new reformulation of foundational dynamics is in the insight it provides into structural aspects of the theory that might have remained obscure had the theory not been reformulated in this novel way. And it is these insights that show the real value of this novel formulation of the theory both for the future development of Newtonian dynamics itself and, again as with the Hamilton equations, for the development of other theories as well.

Finally we shall take up an additional mode of reformulating the fundamental laws of the theory, namely the method of Poisson brackets. Everything said about the place played by Hamilton–Jacobi theory in the development of dynamics can be repeated here. Once again we have a reformulation of existing foundations of some limited value for problem solving, but of immense importance in revealing otherwise hidden structural aspects of the theory and of leading to further conceptual understanding of Newtonian dynamics and of successor theories as well.

15.1 CANONICAL TRANSFORMATIONS

By the nineteenth century it was clear that systems could be described using a variety of different coordinate systems. One could use, for example, coordinates determined by a fixed set of Cartesian coordinates in space. Or, if it was convenient, as it certainly is when doing orbital dynamics, one could use polar coordinates instead. Alternatively, one could use a variety of coordinates not fixed in space but related to the moving body itself, such as the Euler angles, which are so useful for studying the rotation of rigid bodies. Again, in the case of constrained systems it was almost always beneficial to choose a set of generalized coordinates determined by the structure of the constraints, which coordinates would be, unlike the normal spatial coordinates, independent of one another even given the constraints.

With the introduction of Hamilton's equations, generalized momentum is taken to be a variable describing the system itself as fundamental and, unlike generalized velocity, not introduced as defined from the generalized coordinate and its time change. So the possibility of finding the most useful and practical generalized coordinates and momenta for solving a particular problem becomes of great interest.

Suppose that a system possesses a symmetry in one of the general coordinates chosen, so that the Hamiltonian, in standard cases the total energy of the system, is independent of that generalized coordinate. It then follows immediately from Hamilton's equations that the time rate of change of the generalized momentum conjugate to that coordinate will be zero, i.e. that the generalized momentum will be a constant of the motion. So finding generalized coordinates that will reveal this symmetry explicitly becomes a valuable tool for generating constants of motion and simplifying the process of solving the full dynamical problem.

Crucially, however, one wants the generalized coordinates and momenta to which one transforms the description to be such that for them the Hamiltonian equations of motion will still hold. These are called "canonical" coordinates, and the transformations from one such set of coordinates to another are called canonical transformations.

The general method for making such transformations is simple and systematic. One has available four sets of variables: the original and the transformed generalized coordinates and the original and transformed generalized momenta. To make a canonical transformation one needs to choose what is called a "generating function." This will be a function of some pair chosen from the four sets noted above. Simple equations will then represent the variables not appearing in the generating function as derivatives of the generating function (or negatives of these) with respect to the variables that do appear in it. Further, one will have a simple equation that expresses the Hamiltonian function as a function of the new variables as the sum of the old Hamiltonian function and the time derivative of the generating function.

There are basically two ways in which canonical transformations are used to solve particular dynamical problems. One approach is to seek a transformation to canonical coordinates and momenta such that some of them do not appear explicitly in the Hamiltonian function. Then the conjugates of these coordinates or momenta will be constants of the motion. Transforming back to the old coordinates and momenta will provide the results showing how these evolve in time and how the motion of the system proceeds. In other cases one can find new variables such that the derivative of the Hamiltonian with respect to the new variable will be a constant. The conjugate quantity will then be simply a linear function of the time. A different tack is to seek a canonical transformation between the values of the canonical coordinates and momenta at a given time and the initial values of these quantities at time zero (which are, of course, constants). Such a transformation will be just the solution of the dynamical problem, that is, a set of functions giving the values of the coordinates and momenta of the system as functions of their initial values and the time.

Poincaré's theory of integral invariants was originally developed to show how various quantities defined in terms of the generalized positions and momenta would remain constant under any canonical transformation. In particular, volume in phase space defined in the most natural way can be shown to remain invariant whatever one's choice of a set of canonical coordinates. The notion of a canonical transformation can also be used to reveal the intimate connection between the Hamiltonian function and the transformation of a system over time. One can take as a canonical transformation the transformation between the generalized coordinates and momenta of a system at one time and the values of these variables at an "infinitesimal" later time. When one asks for the generating function for this canonical transformation, it turns out to be simply the Hamiltonian function of the system.

For our purposes the notion of canonical variables and the canonical transformations between them will be used only to clarify what is going on in Hamilton–Jacobi theory, to which we now turn.

15.2 HAMILTON–JACOBI THEORY

Hamilton's work in dynamics included more than the discovery of the dynamical equations discussed in the preceding chapter. These, indeed, were found in the course of his broader, overall program. Hamilton's initial work in physics was in the field of optics. It was in seeing the ways in which the two disciplines of optics and dynamics could each provide resources for deeper understanding of the other discipline that Hamilton made many of his most important discoveries. Here the close formal analogy between the least-time principle in optics deriving from Hero and Fermat and the least-action principle stemming from the work of Maupertuis, Euler and Lagrange was the crucial key to how the formal devices developed in one area of physics could be applied to illuminate the other.

Hamilton first seeks a general principle of geometrical optics that will parallel the elegance of the now known general principles of dynamics. Suppose a medium is characterized by an index of refraction that varies from place to place. The light-ray paths through can be found by taking them to be the paths that are the extremals of an "action" function, where the action is taken to be the integral of the index-of-refraction function with respect to distance in the medium. There will be surfaces of "constant action" in the medium and these will be such that all of the light-ray paths will be always orthogonal to these surfaces. Hamilton's formal expression of geometrical optics is indifferent to whether the true nature of light has a projectile nature, as Newton thought, or a wave nature, as Huyghens posited. But if one accepts the wave theory, as was becoming inevitable at the time of Hamilton's work, it is clear that there is an inverse relationship between the speed of light through a medium and the rate at which Hamilton's action is changing along the paths of the light rays, since in the wave theory the speed of light is slower the higher the index of refraction of the medium.

Having illuminated optics by a mechanical analogy, Hamilton then turns to mechanics and casts light on it through analogy with optics. His original work was much improved by C. Jacobi, the resulting theory usually being referred to as the Hamilton–Jacobi approach to dynamics. What Hamilton and Jacobi showed was that a complete solution to a mechanical problem could be found by solving a single partial differential equation for a function of the dynamical variables, usually expressed as S and called "Hamilton's principle function." What S turns out to be is just the action calculated along a dynamical trajectory up to a certain point in the evolution of the system. If we have a dynamical system, characterized by a Hamiltonian function, one can represent the evolution of that kind of system from each possible set of its possible initial generalized coordinate values by now representing the state of a system in a multi-dimensional "configuration space" that has as many dimensions as there are generalized coordinates characterizing the system.

Surfaces in configuration space representing specific values of this S function will be just like wave surfaces of light that propagate in a medium. The relevant "index of refraction" is determined by the form of the potential-energy function of the system. If one is dealing with a single-particle system the configuration space reduces to ordinary physical space, and the "wave velocity" of the surface will be inversely proportional to the spatial velocity of the particle being described at that point of its motion! In nice cases, where the Hamiltonian doesn't involve the time explicitly and is just the total energy of the system, another related function, W, called "Hamilton's characteristic function," can be defined. Its "level surfaces" in configuration space are fixed, and the motion of the level surfaces of the S function is such that a surface for a given S value will coincide at different times with successive W surfaces. In close analogy with the optical case, the system trajectories in configuration space will be "rays" that pierce the S-function surfaces at right angles.

Hamilton (or, rather, Hamilton improved by Jacobi) is able to show that the S function is the solution of a single first-order partial differential equation. Solving that equation provides a complete solution to the dynamical problem. So this single partial differential equation is equivalent to the two first-order differential equations, Hamilton's equations, of the preceding chapter. Finding solutions to partial differential equations is usually harder than dealing with ordinary differential equations, but when the Hamiltonian is separable in the generalized variables of which it is a function, when, that is, the Hamiltonian can be written as a product of functions of the individual variables, it becomes possible to solve the Hamilton–Jacobi equation by evaluating simple integrals. In other cases one can extract information about a system from its Hamilton–Jacobi equation, such as the frequencies of periodic or conditionally periodic systems, without actually completely solving the Hamilton–Jacobi equation. This leads to a method that is quite useful for dealing with problems in celestial dynamics, employing so-called action-angle variables that generalize the notion of the description of a particle in a circular orbit by its fixed angular momentum and its angle from a given line in its plane of motion that increases at a constant rate.

In terms of the canonical transformations discussed in the first section of this chapter the Hamilton principle function and the Hamilton characteristic function turn out to have simple interpretations. The former is the generating function that transforms the problem to a new set of generalized coordinates and momenta in which the Hamiltonian function is identically zero and in which all of the new coordinates and momenta are constants. The latter is the generating function that transforms the Hamiltonian function of the new variables into a constant function and where all of the new generalized momenta are constants of the motion.

Hamilton makes no "metaphysical" claims for his results. Nor is there any claim that the results obtained show some deep *physical* connection between the behavior of light and the behavior of dynamical systems. Indeed, Hamilton emphasizes the "formal" nature of the results. It is abundantly clear that the close analogy between geometrical optics in physical space and dynamics viewed in the framework of configuration space is grounded in the fact that both the optical and the dynamical theories can be based upon extremal principles: in the former theory the least-time principle and in the latter the least-action principle (supplemented by the conservation of energy).

The new framework for dynamics provides no increase in the theory's generality or in its ability to explain dynamical phenomena in some profound sense. It is, rather, one more "recasting" of the existing theory into a format that reveals structural elements of the theory that would not be apparent without the new formal developments. The new formalism is of some help in problem solving. But its major importance is in enlightening us as to the deep structure of the theory and, in its future, in providing rich resources for relating classical dynamics to one of its successor theories, quantum theory.

15.3 POISSON BRACKETS

Poisson initiated the discussion of one more way of reformulating the fundamental dynamical principles. Here the "bracket" notation he introduced allowed a highly symmetric treatment of the generalized positions and momenta of the system. The equations of motion are formulated in terms of a notation that brings out the implicit algebraic structure in the Hamiltonian equations of motion. Once more the Hamiltonian function plays its role as the "generator" of the time evolution of a dynamical system. The time rate of change of any generalized coordinate or momentum was just its bracket with the Hamiltonian function; and the time rate of change of any function of the canonical variables and the time itself was just the bracket of that function added to the explicit rate of its change with respect to time itself.

The bracket notation does not play much of a role in aiding in the solution of problems in classical mechanics. One consequence of it is sometimes of help, since it is easily shown that the bracket of two functions that are constants of motion is also a constant of motion. This can sometimes be used to discover new constants from old. Alas, it usually just leads to new, trivial, functions of the existing constants of motion.

The most important contribution of the bracket notation to the understanding of the theory was its revelation of one more "hidden structure" implicit in the theory. This combinatorial structure can be given a simple formal characterization. That such a structure exists is certainly not apparent from the other formal renditions of the theory. Here again a reformulation of the theory obtained by a simple, albeit deep, mathematical transformation has opened for inspection a structural aspect to the theory that was only implicit in the theory's previous renditions.

This algebraic structure will later function in the twentieth-century projects of placing the dynamical theory's fundamental elements into the forms made available by the abstract mathematics of the later period.

But the most important role the discovery of the bracket notation played was one that went beyond the understanding of the classical dynamical theory itself. It is in the early stages of trying to understand what the new theory that would characterize the quantum nature of the world would have to be like that the bracket formalization of dynamics played its most significant role. For it is this curious, symmetric and algebraically revealing formalization of the classical theory that provides the most direct route for exposing what must be the core formal elements of the new quantum theory.

SUGGESTED READING

An outline of the history of these reformulations of dynamics can be found in Part IV, Chapters V and VI of Dugas (1988). Chapters 8 and 9 of Goldstein (1950) provide a very clear introduction to the mathematical details of canonical transformation, the bracket formulation and Hamilton–Jacobi theory. Chapter XI of Whittaker (1937) covers Hamilton–Jacobi theory in detail.

CHAPTER 16

The search for new foundations

The nineteenth-century reconstructions of dynamics we have just surveyed were certainly not motivated by philosophical reflections on the traditional foundations of the theory. Nor were they formulated in response to any felt disquietude about the traditional theory. Hamilton's equations arise out of an understanding, developing at the time, that a single second-degree differential equation could be replaced by a pair of coupled first-degree equations. Hamilton–Jacobi theory has a less purely mathematical motivation, in that it followed from a deep understanding of the degree to which the formalisms of geometrical optics and of dynamics bore interesting parallels to one another. Here the primary inspiration comes from the understanding that the Principle of Least Time in the former theory and the Principle of Least Action in the latter were sufficiently similar that other formal similarities, such as advancing wave fronts and trajectories as rays orthogonal to these, might be found as well. The bracket formulation of dynamics is, once again, a purely formal manipulation of the theory.

Each of these reconstructions, as we have seen, has manifold consequences. They do play some role in extending the ability of the theory to be applied to difficult special cases. But they also provide deeper understandings of the hidden internal structures of the theory, as in, for example, the realization of the fascinating algebraic structure among the generalized variables revealed by the bracket notation or the "wave-front" structure in configuration space revealed by Hamilton–Jacobi theory. And they also provide just the resources that will later be needed in going beyond classical dynamics to newer theories, as in the application of Hamilton–Jacobi theory to the foundations of the wave-theory version of quantum mechanics and the application of the bracket notations in the formalizing of the matrix version of quantum theory.

But these reconstructions are not motivated by, and do not directly contribute to the resolution of, the long-standing philosophical puzzlement about dynamics.

Philosophical discontent with the standard theory remained much the same from the time of the *Principia* onward. To some, the most unacceptable part of the theory was the necessity claimed by Newton of tolerating a substantival space to serve as the reference frame for "real" or "absolute," as opposed to merely relative, motions. Curiously, Newton's claim of the necessity for an absolute standard of temporal congruence received far less attention. Others objected to the positing of "force" as an ontological category, although the older talk of such posits as reviving Aristotelian Scholastism's "occult qualities" had by and large vanished. Even the positing of "mass" as an intrinsic feature of matter remained disturbing to some. The characterization of gravity as acting at a distance raised the hackles of the later critics, as it did those of the Cartesians, and disturbed even Newton himself.

Finally, there was the ongoing controversy concerning the epistemic and modal status of the basic principles: Was dynamics a rational science founded upon principles that could be known a priori to pure reason? Or was it a discipline founded upon generalization from observational and experimental experience? Were the basic principles necessarily true, or were they merely contingent?

Through the eighteenth century little progress was made on these issues. Daniel Bernoulli and Euler worried about whether the connection of force to acceleration was a priori or empirical, necessary or contingent. D'Alembert, skeptical of the very legitimacy of the notion of force, framed his version of the dynamical generalization of the virtual-work principle in terms of "motions," eschewing the force concept altogether. In the process he made his statement of the principle harder to understand than its later rendition in the Lagrange form. On the other hand, Boscovich took force to be a legitimate primitive concept both in the theory and in metaphysics in general.

The great idealist philosopher Berkeley railed against the notion of absolute space, but contributed nothing new to our understanding of how to do without it and preserve the dynamical theory.

As the nineteenth century wore on there began to be contributions to the philosophical–physical issues that were of more serious weight. The work of a number of scientists pursuing foundational issues in dynamics led eventually to the work of Hertz, and especially, the work of Mach. It continued beyond them in ways that contributed to the further clarifications obtained in the succeeding century.

It is important to recognize that these foundational explorations into the roots of dynamics took place in a broader scientific context. Various theories of matter and of light, of electricity and magnetism, of heat and temperature were all "in the air" at the time. As a consequence the foundational explorations into dynamics often were carried out by scientists whose other commitments – for or against atomism, for or against aether theories of radiation, for or against mechanical theories of heat – often influenced their views in dynamics itself.

Let us look at how these themes developed.

16.1 A SAMPLER OF PROPOSALS

One thread of continuing importance was that which often referred back to d'Alembert. Mass, and particularly, force, were to be avoided as primitive concepts in the theory. Only the motions of matter were to be the basic quantities that should enter into the fundamental laws. Along with this denial of a fundamental ontological place for mass and force there often went an insistence upon the experimental nature of dynamics. Although dynamics could obtain a "rational" form in which all of the dynamical truths followed from some basic set of axiomatic assertions, these axioms themselves had no a-prioristic basis. They could only be inferred by a process of abstraction and induction from observational truths about the world.

In the late eighteenth century Lazare Carnot maintained both these themes. His work, although it stimulated many in the future, was far from ideal in clarity. Much of it, beyond the statement of the methodological themes, consisted in applying versions of the d'Alembert method in its "motions" (as opposed to "forces") version to phenomena.

Barré de Saint-Venant takes up the d'Alembert–Carnot program in the middle of the nineteenth century. "Force" as a basic entity can be conceived of as nothing but an "occult or metaphysical" intrusion into our theory. Only times and places, positions and motions, are the primitive objects with which dynamics is to deal. If we collide identical objects, the motion gained by the one is that lost by the other. If the objects are not identical, the changes of velocity are in constant proportion in any such collision of the two, and it is simply that number that constitutes the relative "mass" of the objects. One can think of objects as made of basic identical atoms if one wishes, and the mass as the number of such basic particles, but the real role of "mass" in the theory is simply that constant ratio our collision experiments reveal.

If we observe a system of points we see that at every moment the particles suffer accelerations toward one another that are directed along the line connecting them and dependent upon their spatial separation but not upon their relative velocities with respect to each other. For identical point objects these accelerations are always equal and opposite to one another.

We can define mass as follows then: "The mass of a body is the ratio of two numbers expressing how many times this body and another body, chosen arbitrarily and always the same, contain parts that, on being separated and colliding with each other two by two, communicate opposite equal velocities to each other."

The bodies communicate opposing accelerations to one another in, say, the situation of mutual gravitational attraction. For simplicity we take those accelerations and multiply them by the again imagined number of basic atoms making the objects up, their masses. This then gives the "forces" to which we say the objects are subjecting one another.

So, "The force, or the attractive or repulsive attraction of a body on another, is a line whose length is the product of the mass of the second body and the mean acceleration of its points toward those of the first; its direction is that of the acceleration."

One shouldn't be misled by the talk of numbers of basic constituents and the like. The real point is that the primitive concepts are to be those of place and time, the changes of place over time and the changes of those velocities, the accelerations. The basic laws are to be the laws governing such changes and how the changes are correlated to the relative positional situations of the objects, whether in collision or in mutual attraction or repulsion. Only after these have been given should notions such as "mass" and "force" be introduced as derivative terms legitimated by the laws discovered.

Ultimately there is to be no association of "mass" with "quantity of matter" in some metaphysical sense. And "force" is to be completely dissociated from any antecedently understood notion of "efficient cause." The route to be traveled by Mach and on into the twentieth century is clear here. As far as I know Saint-Venant has nothing to say about the mysterious matter of the privileged inertial reference frames.

F. Reech presents a foundational perspective diametrically opposed to that of Saint-Venant. For Reech force is foundational, as it was for Euler. He ties the notion of force primarily to its use in statics. Whereas Saint-Venant takes the dynamic measure of force, acceleration multiplied by mass, as definitional of force, for Reech force is fixed in its meaning through the constitutive equations that relate the amount of force present to some feature of the force-generating device. And for Reech real forces are basically those generated by the pushes and pulls of material things on each other. As we shall see, his view of the "mysteriously acting forces" such as gravity is quite different from his view concerning ordinary pushes and pulls.

Indeed, Reech's program is sometimes called "the School of the Thread," since he starts by considering the acceleration introduced to an object by its attachment to a stretched ideal thread that obeys a Hooke law with force proportional to elongation.

Going back to Huyghens, he imagines the object's accelerations before and after the thread pulling it is cut. While attached to the string the object generates a "force of inertia" that balances the tension of the string, a tension that is proportional to the elongation of the thread. Let that force be $F = mf_1$. Now the string is cut and the acceleration of the object changes. Let the change of acceleration be f_2. Is f_1 equal to f_2?

Reech takes it as evident that F was proportional to the mass. He claims that the direction of F and f_2 must be the same since F could not effect motion perpendicular to the thread. Then we might still have $F = mf_2\Psi(f_2, \ldots)$, where the other variables in Ψ might include the place and time of the event. But Reech believes that observation shows us that place, time and velocity of the particle are irrelevant to the change in acceleration. This is a clear recognition of how fundamental the symmetries of spatial and temporal translation and velocity boosts are. So $F = mf_2\Psi(f_2)$. But if we then assume that the effect of each imposed force on the accelerations must be independent of whatever other forces have been imposed then we have $F = mf_2$. In other words the effect of a static force will be to induce an acceleration equal to the force divided by the mass of the particle and in the direction of the force. So here we have an attempt at a derivation of the Second Law as an abstraction and inductive inference from the basic observational facts and from some basic presuppositions as well.

But what if the particle is "free," that is, subject to no push–pull forces from other material things that actually are contiguous to it? Well, if the experiment is done near the surface of the Earth, the particle still is subject to an acceleration, g, the gravitational acceleration in the vicinity of the Earth's surface. Reech is trying very hard not to invoke a special law of inertia or a set of privileged reference frames to which motion must be referred. So his first move is to suggest that we could just take the parabolic motions generated as the "natural" motions of an object not subject to any force.

But the existence of electric and magnetic forces as additional "mysterious" forces stands in the way of this move. Instead, Reech suggests, we must decompose the acceleration of an object not subject to any "effective" force (such as the pull of the thread) into two parts, d' and d'', where md' is the force due to the mysterious agents. Take the effective force F and add to it md' – the force that is due to the mysterious actors. This gives a total force. Then, for simplicity and by convention, take the motion of an object not suffering any total force to be rectilinear and uniform. This sets d'' to zero, since $F + md'$ must equal $m(a - d'')$, where a is the acceleration before the thread is cut. And when the total force is zero we take the motion to be unaccelerated, giving $a = 0$, and, hence, d'' is zero. The idea here seems to be that the law of inertia is just a convention on our part. We simply adjust our ideas of the "mysterious" forces so that we take them in conjunction with effective forces to sum to zero when the motion of the particle is uniform in a straight line.

The crucial aspect of Reech is that his fundamental law is that the effective force is measured by taking the mass of the object multiplied by the change in the acceleration of the object from when the effective force is applied to when it is removed. This law is true even in a non-inertial frame; indeed, it remains invariant between any two frames that are in continuous motion relative to one another, even if they are accelerated frames.

Reech's work is continued by J. Andrade. Reech seems to be presupposing a Newtonian absolute time scale. Andrade thinks that can be dealt with in the following way. Suppose a static force generated by some constant process, such as the force generated at each time by the same elongation of a similarly constructed thread, produces different changes in the acceleration of an object when the experiment is performed at different times. We could then rescale our clocks to force these acceleration changes to be the same at each time. When we did we would obtain an "absolute" rate of the lapse of time that would be Newton's absolute time scale.

Has the Reech–Andrade program eliminated the need for positing some privileged reference frame? Apparently not, for Andrade himself emphasizes the fact that the inertial frames reappear in this reformulation of dynamics in the form of those reference frames such that the forces particles exert upon one another will be equal and opposite only when the accelerations are referred to these distinguished reference frames. But the ideas of Reech and Andrade – of forces as measured by changes in acceleration, of motion under gravity to be construed as natural motion and of absolute time as the time scale in which motions are most simply described – will reappear both in later work on classical dynamics and, in much changed guises, in general relativity.

H. Helmholtz's contribution to the debates was less significant. He was greatly impressed by the conservation of energy. In the case of particles moving under mutual attraction, where the forces are conservative, kinetic energy generated as force integrated over path length, already called potential energy by J. Rankine, is lost. Helmholtz was deeply moved by the experiments then going on that showed a universal status for the conservation of energy even in the processes where it previously seemed not to hold, such as in inelastic collisions and in cases where there is friction, so long as heat energy was taken into account. Could not all dynamics, he speculated, be reduced to the conservation of energy if the inner structure of matter turned out to be that of microscopic constituents acting upon one another by means of conservative forces? But, of course, conservation of energy by itself will not provide a unitary grounding for all of dynamics.

16.2 HERTZ'S NEO-CARTESIANISM

Heinrich Hertz, who was distinguished as an experimental physicist on account of his demonstration of the reality of electromagnetic radiation, essayed a thoroughgoing reformulation of classical dynamics. He combined this with a view about the very nature of scientific theory that has resonated in various ways since he first proposed it. His actual reconstruction of dynamics, one that combined a new fundamental principle with a very imaginative reconstruction of the ontology presupposed by the theory, remained, however, an oddity in the history of the attempts at finally dealing with the theory's persistent conceptual anomalies.

Hertz takes it that the ultimate purpose of a theory is to provide us with an adequate rule for predicting new observable consequences of known observable facts. Theories are, he says, "images." Perhaps "representations" would be a better translation. We construct these images to do the predictive job. The images must be logically consistent. And they must be true to the observed correlations among the observable facts. But we desire them also to be "convenient" or "appropriate" (*zweckmäßig*). Whereas logical consistency and truth to the observables are objective criteria of a theory's acceptability, convenience is a matter largely of taste. There can, then, be multiple versions of a theory, with each version consistent and experimentally sound, but with the different versions providing distinct images, each taken by some as most "convenient." Images should have a rich enough structure that each element in the reality to be described is accounted for. But they should be "simple" as well. Here simplicity is taken to mean that there should be no excess structure in the image. There should be no features of it that function as "idle wheels" in the mechanism, as does, for example, absolute rest in Newton's theory.

What is wrong with the traditional formulation of dynamics? Not logical inconsistency or empirical inadequacy, although, Hertz says, there is no reason to believe that the theory will not be found empirically inadequate at some future time. But the theory misleads us by its peculiar conceptual way of dealing with the phenomena. The problem rests primarily with the concept of force. Hertz gives the following example of the kinds of confusion the notion of force can result in. A whirled rock on the end of a string is held by a force of tension in the string. The Third Law demands an equal and opposite force. But the "centrifugal force" of the rock is nothing but its inertial tendency to travel in a straight line, not properly a force at all.

Hertz believes that the existing formulation of dynamics is too broad, allowing for kinds of forces and kinds of constraints never met with in nature. Perhaps it is the excessive broadness of this "image" that leads to trouble. Perhaps the fault lies in the image having too many inessential characteristics. In particular the notion of force is otiose. Forces may seem real enough when it is the weight of an object or the force of our arm in throwing something that is in question, but the existing formulation of dynamics leads us to posit too many phantom action-at-a-distance forces. The standard image then is unacceptable. It leaves individual concepts unclear (force) and it allows the possibility of motions that are never encountered in nature.

What of an approach founded on the notion of energy? Here Hertz says the problem is that many energeticists (Helmholtz) never make it clear just what the theory formulation is.

One possibility would be to adopt Hamilton's principle, that the integral of the difference between kinetic and potential energy over the dynamical path is extremal, as fundamental. But would this even be a correct version of the theory? Hertz argues that the principle holds only in special dynamical cases. The problem is that there exist constraints, such as a sphere rolling on a plane without slipping, that cannot be dealt with in this formalism. This is a problem not only for Hamilton's principle, but for any attempt at formulating the theory with an integral principle as foundational. Here Hertz is too quick. It is true that one version of Hamilton's principle will not function properly in the case of systems subject to non-holonomic constraints, constraints that are not integrable and are such that the constraint cannot be used to lower the number of degrees of freedom of the system. But shortly after Hertz's work was published O. Hölder was able to show that a subtly different interpretation of the variational method allowed Hamilton's principle to be applied to cases such as the sphere rolling on the plane as well as to cases where the constraints were holonomic.

Furthermore, Hertz asserts, the energetic version of the theory requires finite potential energies, a condition that might not hold in an infinite Universe. And the notion of minimizing the integral in question has no simple physical meaning. Finally, it is unacceptable to put a principle that seems to invoke final causes or teleology at the foundation of physics.

What then is the alternative that Hertz will offer us? Hertz wishes to use only the concepts of space, time and mass. The concept of force is to play no primitive role, although, as we shall see, it can later be introduced as a derivative notion. But we must postulate something beyond the ordinary visible matter of the world if we are to do justice to such dynamical phenomena as planetary motion, since we have rejected any notion of forces acting at a distance. Invisible matter and its motions must be posited as well. Here Hertz is working in an environment where the postulation of an all-pervasive aether as the medium of transmission of electromagnetic waves is a common belief, and where even models of atoms as vortices in the aether are in prominent use.

The dynamics of visible matter is governed by the fact that its motion is constrained by the presence of the invisible matter. Here the contrast with those who think of dynamics from an atomistic perspective is striking. Consider a constrained dynamical system, say a bead constrained to slide frictionlessly on a wire. For most atomists macroscopic constraints are not fundamental. What is really going on is a complex set of forces acting between the atoms of the wire and the atoms of the bead. For practical purposes these can be summarized as "forces of constraint" holding the bead to motions along the wire. And, as a long tradition from James Bernoulli through Lagrange had shown, the dynamics of the bead can be characterized without ever calculating what those constraint forces amount to. For the geometric constraint on the motion of the bead can be dealt with, and the dynamics of the bead predicted, without ever calculating the magnitudes of the constraint forces.

But for Hertz it is a different story altogether. For him there is only what would otherwise be free motion with uniform speed along a straight line, except to the degree to which the motion of the system in question is constrained by the existence of other masses, visible or invisible, which serve to constrain the motion of the mass in question. It is constraints that are fundamental. If any notion of force is to be introduced, it must be a notion derived from the fundamental laws governing constrained motion. Not all conceivable constraints will be allowed into the theory. That would be to make a mistake similar to that of the "force" version of the theory that allowed all kinds of possible forces that are never encountered in nature. We could, for example, imagine a system constrained to always maintain a fixed angle with some plane. But such constraints will not be countenanced. Only constraints that are mathematically representable by homogeneous linear equations between first derivatives of the motion will be allowed. Such constraints include holonomic constraints (the term is due to Hertz), in which case they are integrable. In that case the constraint will be expressed as a function on some of the coordinate variables being constant. But Hertz will also allow constraints such as those imposed on a disk or sphere rolling without slipping on a plane, which are not so integrable and are therefore, by definition, non-holonomic.

Hertz develops a way of looking at the kinematics of systems that is itself of great historical interest. The method will use much of the mathematics of general *n*-dimensional curved spaces that comes from Gauss and Riemann. But, interestingly, Hertz was quite sure that such geometries had nothing to do with real, physical space. About that, he was convinced that Euclidean three space was all there was to the matter. The mathematics is applied not to physical space but to the configuration space of a system. Here it is the path in that space representing the simultaneous evolution in physical space of all of the masses in the physical system that is at issue. Such a path has a curvature given as a function of the accelerations of the individual masses of the physical system. And a notion of a path of least curvature can easily be defined.

A single fundamental law then governs all of dynamics: Every free system persists in its state of rest or uniform motion in a straightest path. Go to the configuration space of a system that is not in interaction with any masses outside the system (i.e. is not constrained by any other masses). The possible path of such a system's representative point in the system's configuration space will be determined by the internal constraints in the system which tie its components geometrically to one another in their motions. One such path will be the one with least curvature at every point. The representative point will traverse that path with constant speed.

Now, if the constraints are holonomic, Hertz acknowledges, such a path will also be one of extremal length beween two points along it when compared with appropriately selected nearby paths. That will be the content of Hamilton's principle or one of the other extremal principles. But not all constraints are holonomic. Furthermore, Hertz's principle is not an integral principle. He is therefore delighted to remind the reader that it does not bring with it all the dubious notions of final causes, teleology, and, perhaps, religion, that had accreted around the standard extremal principles. In addition, this basic law, and the geometric picture of dynamics in configuration space that goes with it, leads easily to the methods of Hamilton–Jacobi theory and to an understanding of its characteristic functions.

This new single dynamical law is then combined with the cosmological picture of a world consisting of visible and invisible masses in motion. The behavior of the visible masses is governed by the single fundamental dynamical law, of course. But how the visible masses behave will depend upon how their motion is constrained by their geometric connections to the invisible masses. We don't know, though, how the invisible masses are distributed or how, exactly, they constrain the visible masses.

Comparison with Descartes' vortex model of the motion of the planets is interesting here. Both theories take the planetary motion to be constrained by the existence of a mass-in-motion coupled to the motion of the planet. In Descartes' case this is the swirling plenum and in Hertz's case it is the invisible aether. Descartes insisted that the medium governing the motion be, like all matter, characterizable simply in geometric terms. In that sense it had to be as "manifest" as the planets themselves. Hertz allows the introduction of mass as a primitive characterizing both visible and invisible matter, and his aether is not open to visual determination. Hertz never specifically insists that the constraining all be done by continuous contact of visible and invisible matter, but he plainly has that in mind. It is not just the fact that forces are introduced in the first image that he rejects, but also the fact that these forces act at a distance which is reprehensible.

It is interesting to reflect here on some general methodological ideas about the legitimacy of concepts in physical theories. There are those who would reject any concept referring to an "unobservable." Hertz plainly isn't one of these. But there is another school that would take as intelligible concepts that refer to the unobservable as long as the terms expressing those concepts are taken to "mean the same thing" as they do when they are used to describe the observables. Here I think is where Hertz stands. He rejects "force" as a legitimate meaningful term in physics, but not simply because it refers to the unobservable. It has nasty metaphysical connotations and it leads to conceptual confusion, as illustrated by the notoriety of "centrifugal force." But spatial, temporal and mass predications are all legitimate. And the predications "mean the same thing" irrespective of whether they are applied to observable or unobservable matter. So the reference to the unobservable cannot be criticized, according to Hertz, as requiring the introduction of new conceptually dubious concepts into physics. Invisible matter is just like ordinary matter – except that it is invisible!

Much of Hertz's effort is taken up in showing how this new "image" and its law can do the job of the other approaches to dynamics. Some of this is merely formal, such as proving that in the case of holonomic constraints the Hertz law leads to Hamilton's principle. The real task, however, is explaining how the new approach can deal with such classic successes of the familiar version of dynamics as its derivation of the laws of celestial motion from the fundamental dynamical laws supplemented by the law of force acting at a distance – the law of universal gravitation.

A good deal of effort goes into showing how certain generic kinds of constraint relation from invisible to visible matter can give us familiar results. These can often be obtained even if the detailed structure of the invisible matter is unknown. For example, if the constraining relationship is adiabatic, with visible and invisible matter not exchanging energy with one another, then conservation of energy will still hold for the visible matter taken by itself.

Much important work goes into showing how an appropriate set of constraining relations imposed on the visible matter by the invisible could result in a situation where the motions of the elements of visible matter would be as if they were being influenced by forces exerted on one another, and in showing that in some cases these forces will be derivable from scalar potentials. The forces, not surprisingly, turn up as Lagrangian undetermined multipliers in the constraint equations. The trick is in showing that there will be appropriate force–acceleration equations that are closed in the motion of the visible matter alone.

The situation is a bit strange. We don't know what the structure of the invisible matter is, or exactly how it serves to constrain the motions of the visible. So what we actually have to do in practice is derive the "force" functions from the observable behavior of the visible matter and go on from there. It is only a matter of principle that we recognize that really there are no such things as forces and that the motions we see are really the result of contiguous constraints on the motion of the visible by the invisible.

Some of the greatest difficulty for Hertz is encountered in dealing with Newton's Third Law. Remember how that law was originally derived for contact processes by the empirical observation of momentum conservation combined with the idea of the contact forces serving as impulses over time to change the momenta of the colliding objects. Remember also, however, how Newton clearly understood that extending the law to matter interacting by forces at a distance was a substantial leap. He argues for the legitimacy of this leap first by a thought experiment. Two masses are connected by a rod. If one attracted the other with a force different from that which the second exerted on the first, there would be spontaneous acceleration of the rod. Then he backs this up by reference to his experiments using floating magnets. Furthermore, the invocation of the Third Law is essential in arguing for the universality of gravitation and for applying the dynamical principles to determine the cosmic motions. Without it one cannot get the result that it is the center of gravity of the solar system around which all bodies, including the Sun, move, rather than, say, the geometric center of the Sun itself.

Hertz can certainly show that, if one has a single free system that is subdivided into two parts, each part constraining the motion of the other, his fundamental law guarantees that any momentum gained by one part will be lost by the other. But that is quite a different matter from showing that there applies for the visible matter by itself a version of the statement that the "forces exerted on one another will be along the line joining the masses and equal in magnitude and opposite in direction" that is needed for duplicating the results of the Newtonian synthesis. To show that, we would need to know much more about exactly how the invisible masses constrain the motions of the visible. He does give an argument that applies to two independent visible masses, say in collision, but has no way of getting Newton's pairwise equal and opposite forces for multiple bodies interacting by action-at-a-distance forces.

Hertz at one point argues that this may be a good thing. For, he says, in the case of the interaction of current elements, at least in the case where the currents are not stationary and the momentum carried by the electromagnetic field must be taken into account in momentum conservation, "it is not quite clear what is meant by opposite." But that seems a peculiar way of avoiding the fact that his system really is unable to duplicate the full Newtonian principles which still apply when all the relevant forces and objects on which they act are taken into account.

On reading Hertz one is puzzled by how little attention is paid to the other notorious puzzle about the Newtonian theory: the existence of privileged reference frames, namely the inertial frames, and of a privileged standard for temporal congruence, namely absolute time. Hertz takes spatial congruence to be fixed by measuring rods and the temporal interval by a "chronometer." He has nothing to say about the origin of the privileged status for the inertial frames or about the origin of "absolute" time.

Indeed, in an extremely insightful critique of Hertz Simon Saunders has noted a deep way in which Hertz's system fails to be a satisfactory substitute for Newton's. Newton went to great lengths to show us how we could determine whether or not a system was in inertial motion or not, and he implicitly offers us a method for determining the absolute lapse of time as well. We shall look at these matters again in some detail when we explore contemporary attempts at building a sound Machian relationist account of dynamics. Once we have postulated a force function in Newton's "image" we can determine, for example, whether or not a system is in rotation, as

Newton's "rocks on the end of a rope" passage in the "Scholium to the Definitions" of the *Principia* is designed to show. And astronomers knew very well in Hertz's time how to calibrate their chronometers correctly. Basically, one took as much of the dynamics of the solar system into account as one could and then sought the right time keeping that would fit the observed motions, with as many perturbations taken into account as one could, into the pattern demanded by the Newtonian dynamical and force laws. This establishes so-called "ephemeris" time.

But, Saunders argues, without knowing the structure of the motions of the invisible masses and the constraints all of the masses, visible and invisible, exert upon one another, Hertz has no way of fixing the inertial frames and the absolute time scale. His only resort is to simply assume the Newtonian structure, including the force law of gravity and the full Third Law, and then go on as in Newton. But for Hertz there is no "principled" way of establishing these laws, especially the full Third Law. These principles in his system are mere "phenomenology" akin to the role played by Kepler's Laws in the Newtonian framework. So, Saunders argues, Hertz has no "principled" means of establishing that which is essential for the correct application of his dynamics to the world: a specification of the inertial frames and of the absolute time scale. Whereas in Newton's "image" the rules for finding the fundamental reference frames and time scales are available in the fundamental laws of the theory, in Hertz's "image" no such rules are forthcoming from the first principles of the theory. They can be discovered only by assuming the "phenomenological" empirical fact that the Newtonian force laws hold for the visible matter as an unexplained consequence of the constraints on it from the invisible matter. But these rules are essential in order for one to be able to apply the fundamental theory to the world.

Hertz's approach to dynamics remained an idiosyncratic and isolated foray in the later history of the theory. For one thing, both the model of radiation as mechanical motion in an all-pervasive aether and the vortex model of atoms soon dropped out of accepted physics. But it was only against this background that Hertz's invisible matter had some kind of antecedent plausibility. In later physics the notion of "constraints" did play an essential role as auxiliary equations needed along with the fundamental dynamical equations in gauge field theories. For example, the principle of conservation of charge is expressed in Maxwell's equations by a constraint. But the kind of constraints Hertz had in mind, namely geometric conditions on related motions that were to replace altogether forces and that were direct generalizations of macroscopic constraints that were familiar from the notion of a rigid body and a continuous fluid, for example, never again appeared as substitutes for forces or their equivalents.

The work of Hertz is fascinating for us, however. The fact was that a whole approach to the foundations of a central theory that had been discarded as not only wrong but utterly wrong-headed, that is to say Descartes' approach with its plenum fluid and celestial motions governed by the contiguous constraint of the fluid's motion on the embedded planets and moons, could be reaffirmed in a subtly transformed version. And this time one could not simply dismiss the view because of its empirically awful predictions, as Newton did the Cartesian picture in the second book of the *Principia*.

The motives behind Hertz's view are certainly curious. Some of the objections to the notion of "force" were misguided. The issue of "centrifugal force" can be completely clarified from the Newtonian perspective. The desire to restrict the allowed conceptual vocabulary to spatial and temporal terms and to a concept of mass has a peculiar feel to it. Most other revisionary approaches demanded a "definition" for "mass" as well as for "force." But for Hertz, as for d'Alembert, it was the "metaphysical" nature of force, the idea of it as "efficient cause," and the puzzle over the status of the fundamental law binding force and acceleration (necessary or contingent, a priori or empirical?) that made it especially desirable to rid the theory of that notion. That, along with his irritation concerning the idea of forces acting at a distance, was what made Hertz desire a dynamics without force. Behind the whole project lie intuitions about the nature of legitimate explanations in physics and the limits of legitimate concepts in physical theories to which we will need to return.

Hertz's general view about the nature of scientific theories was influential among later philosophers of science. But it is far from clear, though, what view one ought to attribute to Hertz. He tells us that "appropriateness" is one of the grounds, above consistency and empirical correctness, by which theories are chosen. His remarks on the ideal theory as having just enough structure to capture all that is necessary and no more structure than is essential are certainly on the mark. But is he a "realist" or "conventionalist" or what with regard to the posits of our theories? His observation that more than one image may be deemed appropriate seems to suggest some kind of conventionalism or instrumentalism with regard to the posits by theories of unobservables. But his actual work in dynamics seems to be that of someone who is truly exploring the possibility that the invisible masses "really" exist and "really" serve to constrain the motion of the visible masses. As we shall see, Hertz is not the only scientist-philosopher of the late nineteenth century whose views on the real nature of our theories are hard to pin down.

16.3 MACH'S RELATIONISM

As a philosopher Ernst Mach was a phenomenalist. We are to take as the fundamental entities of the world the immediate contents of sensation, for these are the only elements directly given to our experience. The postulation of other entities or features of the world is to be taken only as the introduction of concepts that allow us to describe the orderly, lawlike, structure we find among the immediate phenomenal world of sensations. Material objects and their perceivable properties and relations are to be generated in this way. Mach is also a convinced empiricist in the sense that the fundamental laws of the world are to be inferred inductively from experience. Any hope of finding such laws as truths of reason or a priori in some other way is out of the question.

Just as one can discuss the relationism of Leibniz without probing the metaphysical depths of his monadology, for our purposes we need not explore Mach's deepest philosophy. With regard to dynamics he was, like Leibniz, a relationist. Talk of the positions of objects with respect to one another, or of the motions of objects relative to one another, was prima facie legitimate. So was talk of the rate at which one clock proceeded as judged by some other observable clock. But to speak of "absolute" positions, velocities or accelerations of objects, relations the objects bore to "space itself" was, for Mach, obviously illegitimate. And to speak of the rate of a clock with respect to time itself was equally absurd. Here the fundamental ground for dismissing such notions was the usual one familiar from relationism throughout the ages. Since space itself is not available to us by observation, and time itself is equally remote from epistemic accessibility, any reference to them, or to any feature requiring their existence, is totally unacceptable. Indeed it is "nonsense."

We are not, though, concerned with the general philosophical debate between relationists and "substantivalists" with regard to space and time. Our interest is in the special path the debate took at the hands of Newton. For it was Newton who introduced the argument from dynamics for both absolute space and absolute time. Inertial motion is dynamically distinguished. But without an absolute time scale, the notion of a body moving with unchanging speed is empty. And without a notion of an absolute spatial reference frame the notion of motion in an unchanging direction is equally empty. Both notions are needed to characterize genuine free, inertial, motion. Any deviation from that motion reveals itself to us by the presence of inertial forces, either those of linear acceleration or those of rotation. How will Mach the relationist deal with these familiar Newtonian arguments?

Leibniz sometimes argued that there really were no motions except inertial motions and sometimes that the forces generated by non-inertial motion could be accounted for by the fact that to put something into a non-inertial state required the application of an applied, external force. Huyghens, although himself emphasizing how absolute rotation really was, tried to characterize rotational motion in terms of the relative velocities of the point masses making up the rotating rigid body. Berkeley denied the existence of absolute space. He refers to the "fixed stars" as our implicit reference frame when we speak, say, of the Earth being in motion, but doesn't offer motion with respect to them as an explanation of the dynamical facts Newton referred to in claiming motion to be absolute. He suggests that in a relative motion the object truly moved is the one whose motion is generated by some force external to it. But this will not really do since an eternally rotating object would still show the inertial effects. It is worth noting that each of Leibniz, Huyghens and Berkeley is aware that Newton's dynamical arguments show that relationism must be somehow supplemented to account for the asymmetric fact that even in some relative motions it is one object and not the other that shows the inertial effects. Neither Kant, who was sometimes a relationist, sometimes a substantivalist, and eventually tried to find something in between, nor Maxwell, who sometimes espoused relationism but was well aware of the force of Newton's arguments, and who did bring to attention one matter of later importance, namely the empirical irrelevance of uniform linear acceleration of all matter, had any serious answer to Newton.

As well as dealing with the problem of the "absolutes" of Newton's theory, the matter almost totally ignored by Hertz, Mach had things to say about the concept of mass. Here he takes up the idea we have seen in Saint-Venant and does add something useful to it. Mass ratios are simply defined by inverse ratios of velocity changes induced in pairwise collisions. Mach points out clearly the need for a consistency condition: the mass ratio of *a* and *c* must be the product of the ratios of *a* to *b* and *b* to *c* for any third mass *b*. If this were not so, then by a closed series of collisions we could arbitrarily increase the velocity of a particle. We see empirically that we cannot do this, indicating that the consistency condition holds. Here was a serious beginning to the discussion of whether the theoretical term "mass" could be defined in terms of some putative observables (relative motions). But only the beginning, for Mach's clever device leaves open the question of whether the values of relative masses can be fixed for any system of particles moving under a system of interactive forces, not just for a simple isolated pair.

When it comes to absolute time Mach really just avoids the issue. He tells us that it only makes sense to speak of the rate of one clock as compared with another, and that it is nonsense to speak of the rate of a clock relative to the rate at which "time itself" flows. But he doesn't add anything to the question of why there is such a "preferred" time scale as that posited as necessary by Newton, nor does he have anything to say that goes beyond Newton on how that rate of "time itself" is to be determined by us and employed by us in empirical dynamics.

Mach's name has become attached to a doctrine regarding the preferred spatial reference frame. On this he has much more to say. But he is cautious to the point of being somewhat evasive about what he really believes. Mach is certainly aware of the need for such preferential frames. What do we know about the special frames? First the experimental facts seem to show us that the "fixed stars" represent a reference frame relative to which rotations have their dynamical effects. If we measure the rotation rate of the Earth dynamically, say by means of a Foucault pendulum, we get the same rate as that measured relative to the stars by observation. Mach refers to results of H. Seelinger that indicate a rotation rate for the "fixed stars" of at most a few seconds of arc per century. Mach is aware that all of this is tentative. He looks toward a future in which it might turn out differently, so the fact that what he called the "fixed stars" are stars in our galaxy, a galaxy which has an observationally determinable absolute rotation, shouldn't be held against him. Indeed, he sometimes talks of the more general notion of all of the mass of the Universe, rather than the observable fixed stars of his time, as being the reference frame he has in mind.

What he says is that we do not have good experimental reason to disbelieve that it is rotation relative to the fixed stars, rather than to "space itself," that is responsible for the dynamical effects noted in Newton's spinning-bucket experiment. We have no reason, he suggests, for denying that, were the bucket left alone and the fixed stars rotated around them, the forces would not appear. Nor do we have good reason for thinking that, were the sides of the bucket increased to vast proportions in the early part of the experiment, when bucket spun and water rested but was in motion relative to the bucket, we would not see concavity in the water's surface. "The principles of mechanics," he says, "can be so conceived, that even for relative rotations centrifugal forces arise."

But what about the expected Newtonian replies? Newtonians argued that even in an otherwise empty Universe there would still be a distinction between a rotating globe and one that did not rotate. The rotating globe, for example, if not perfectly rigid, would show oblateness by bulging at the equator. C. Neumann had argued that, if we have such a rotating globe in our Universe and imagine the fixed stars obliterated, we have no reason to suspect that the bulge in the globe would vanish. As far back as *De Gravitatione* Newton had argued that, had God left the Earth alone and rotated the heavens about it, this would generate enormous forces tending to disrupt the heavens, not the usual evidence of inertial forces seen in a spinning Earth.

But, Mach says, "The Universe is not *twice* given, with an Earth at rest and an Earth in motion; but only *once*, with its *relative* motions alone determinable. It is, accordingly, not permitted us to say how things would be if the Earth did not rotate." The Newtonian objections are "thought experiments." Such talk about how things "would behave" were circumstances other than they are is legitimate if it is only "small" changes in how things are that we have in mind. But no inferences from how things are to how they would be were major structural features of the Universe changed are allowable. The contemplated obliteration of all the fixed stars is just the sort of change whose outcome we cannot infer.

There are two ways of interpreting these remarks, although it is dubious that Mach ever realized that. One is to say that we have no good grounds for accepting the truth or falsity of such grand counterfactuals. The other is to deny that such grand counterfactuals have any determinate truth value at all. My guess is that Mach intended the former interpretation but would be even happier if the latter were suggested to him. That would fit with his view of laws of nature as being nothing over and above the economical description of the general ways in which things are in the actual world.

As we noted, Mach is very cautious, only saying that we do not have good reason to say that "absolute rotation" might just be rotation relative to the fixed stars, and suggesting that a dynamics in which relative rotations generated inertial forces could be formulated. But it is pretty clear that he believes the Newtonian theory wrong and some relationist account correct. Implicit here and elsewhere in his work is the idea that there are forces generated between relatively accelerated masses and that these forces are highly mass-dependent and highly distance-independent. The relative rotation of the water in the bucket or of the Earth with respect to the distant but massive matter of the Universe is the source of the inertial forces, not the rotation of these test objects with respect to space itself.

But Mach offers no such new dynamical theory. His work served, and continues to serve, as a stimulus to those who wish to go further and construct an empirically acceptable relationist mechanics – with striking and surprising results, as we shall see. The idea that the "average smeared-out mass of the Universe" could itself serve as a substitute for absolute space remained a tantalizing one. Of course, Mach is on much firmer ground when he, like so many others, points out the peculiarity of Newton's own particular account with its empirically totally irrelevant absolute position and absolute uniform velocity.

16.4 FURTHER FOUNDATIONAL REFLECTIONS

G. Kirchhoff offered a formalization of mechanics that, for the most part, merely organized the well-known basic principles in their Newtonian format.

Conceptually, Kirchhoff is important for his suggestion that the notion of force was simply to be defined as the acceleration induced on a unit mass. He is aware, however, that something is not quite right here. For, if a system is subject to a number of forces of different kinds, these cannot be defined in his simple manner. Only the resultant, total, force is evidenced in the actual acceleration of the object. Such considerations play a vital role later in the more rigorous formalizations of dynamics that we shall explore in Chapter 20.

Henri Poincaré's contributions to the science of dynamics at the end of the nineteenth century are second to none. Poincaré was the inventor of the qualitative and topological methods in dynamics that later brought important foundational insights to many of the most important fields of application of mechanics. Modern celestial mechanics, chaos theory and statistical mechanics all owe vast debts to Poincaré.

In two essays printed in his *Science and Hypothesis*, "The Classical Mechanics" and "Relative and Absolute Motion," Poincaré also offered an attack on a number of the familiar, outstanding philosophical questions in the foundations of dynamics. Described by Dugas as "rightly famous" and "deeply penetrating," the essays are thought-provoking pieces rather than completed formal accounts that give full-fledged solutions to the puzzles attacked. They are replete with insights, but leave the fundamental questions open for fuller and deeper attacks in later years.

These essays followed those in which Poincaré took up the issue of the epistemology of space and time. There his fundamental argument was that only local coincidences of material objects and events were within the realm of the observationally determinable, with space and time themselves being "non-intuitive." With such distant relationships as the spatial congruence of separated rods or the equality of intervals ticked off on a clock at spatial or temporal separations outside the realm of perceptual determination, it was always possible for a scientist to choose a geometry for spacetime or a standard for time lapse at will. Geometry then was a matter of "convention," and one could do one's physics within a Euclidean or non-Euclidean framework as one chose. Poincaré's critique also anticipated Einstein's famous critical exploration of distant simultaneity that later provided a basic foundation for the special theory of relativity.

In the essays on mechanics Poincaré is once again determined to show us that the basic posits of our theory are true only as a matter of convention. In his essays on space and time Poincaré also argued that, since the geometric axioms were conventionally adopted, they had nothing to fear in the way of possible refutation by any future observation or experiment. He argues that the same is true of the fundamental laws of classical dynamics. In both cases, of course, this particular view of Poincaré's soon became an embarrassment as the relativistic theories adopted both non-Euclidean geometries and novel dynamical postulates and quantum mechanics rejected the deepest presuppositions of the classical dynamical account of the world.

The first essay of Poincaré is devoted to the two inter-locking topics of the place in the theory of the concepts of mass and force, and the epistemic status of the fundamental dynamical laws. The treatises of mechanics, Poincaré tells us, don't tell us what is purely mathematical, what conventional and what experimental in the basic propositions of the theory. Further, they presuppose absolute space, even though there is not only no such thing, but also the very Euclidean nature of space and the simultaneity of distant events or the equality of distinct time periods are all matters of convention. For the purpose of these essays on mechanics, though, Poincaré will ignore the last remarks and just assume a Euclidean space and absolute time.

The first law, the law of inertial motion, is certainly not a priori nor "obvious to the rational mind." So how is it verified? Take the law in what Poincaré claims to be its more general form: The acceleration of a body depends only on its relative position and relative velocity with respect to nearby bodies. Why do we take it that dynamics is formulated in the manner using a second-degree differential equation instead of, say, with a first-degree equation in which it is velocity that is so correlated with the relations to nearby objects or a third-degree equation in which it is change of acceleration and not of velocity that is so correlated? If we only saw the planets moving in unperturbed circular orbits about the Sun we would opt for the first alternative. Only when a perturbation made the orbits elliptical would we be forced into our standard choice.

Is this general principle verified by experience? Newton certainly claimed so, referring to the work of Galileo and of Kepler for his justification. In astronomy, for it to be incorrect would require curious coincidences that mislead us. But what about in the non-astronomical situation? According to Poincaré we have no doubts that the law will ever be changed. And we are justified in thinking this. But how can that be if the law is itself not a priori? In order to truly test the law we would need to repeat the entire evolution of the Universe over a period of time, beginning it more than once with exactly the same initial positions and velocities of all its particles, to see whether the evolution would be identical as the law demands. But of course we cannot do this. If it ever seemed to us that the law was violated we could always have recourse to the postulation of hidden interfering bodies whose relations to the test system in question we had ignored. So the law is actually a matter of convention.

When we get to the "law of acceleration," $F = ma$, things get even more complicated. To test the law we would need to measure independently three quantities: F, m and a. We assume a is measurable, but what about F and m? "We do not even know what they are." Newton's definition of mass is notoriously defective and Kirchhoff resorts to the expedient of just defining F as ma. Worrying about whether force is the "cause" of acceleration is just metaphysics. For science it is simply the measurement of force that is crucial.

But how do we determine the equality of two forces? Is this the case when a is the same for the same m? But that requires already knowing when the masses are equal. Is this the case when forces in opposite directions at a point are in equilibrium? But most forces whose equality or inequality we are interested in are not so disposed relative to one another, and forces cannot be detached and reattached assuming that their magnitudes remain unchanged. It is impossible for us to say what acceleration a force "would" give to a body were it applied to it, or whether two forces would balance "were" they opposed at a point. Experiments with dynamometers, say, just assume such laws as that of the equality of action and reaction in their application. Insofar as Kirchhoff is concerned, he needs a prior notion of the magnitude of m in order to get the magnitude of F even given his definition and the magnitude of a. He can only get the ratio of forces applied to the same body at different times (assuming an unchanging mass).

Well, use the actions of A and B on one another (the accelerations they induce on one another) to define their mass ratio as the inverse of their induced acceleration ratio. The *constancy* of this ratio would then be an experimental fact. This would be satisfactory if we could assume that A and B constitute an isolated system. But bodies are never truly isolated. In order to get this procedure under way, then, we would need to decompose the accelerations bodies receive into those due to different interactions, and we would need to assume that all of these are independent of one another. In astronomy we can move in this direction by assuming that the forces are all central forces. Actually we would be measuring not inertial mass but gravitational attractive mass. But what right do we have to make such assumptions about the force laws? And, if we reject the posit of central forces, along with that goes an experimental basis for the action–reaction principle as well!

So what is left? We could assume the uniform and rectilinear motion of the center of mass of an isolated system (the First Law) and ask what assignment of masses to the components of the system makes this come out true. But, again, systems are never truly isolated. In order to start to independently define the masses in this way, we would have to use the whole Universe as our system – and this we cannot do.

We could, if we wanted to, always assign novel masses to the bits of the world and still not refute the basic dynamical laws. All we would need to do would be to modify the force assignments and laws. But then our system wouldn't be so "simple." Hertz thought that the mechanical principles might, in time, be refuted by experiment, but he was wrong. The principle of acceleration, $F = ma$, and that of the equality of action and reaction are "definitions." Why then study dynamics?

Well, there are systems that are "nearly isolated." For these we study the relative motions of the components – and not such things as the motion of the center of mass with respect to the rest of the Universe. These systems are such that they can be described as nearly in conformity with the laws when we make appropriate assignments of masses and forces. That is an experimental fact. And so, Poincaré concludes, experiment can provide a "basis" for the laws of mechanics, but it can never refute them.

Kirchhoff is accused of the mathematician's tendency to nominalism when he wants to define force. It needs no definition because it is "primitive, irreducible and indefinable." We have, of course, our "intuition" of it in the experience of effort, but that is insufficient and useless in mechanics, where it is only the measurement of force that plays a role. These subjective ideas are just the vestiges of anthropomorphic thought. Further, the work of Reech and Andrade, the "School of the Thread," is of no real help at all. Even when real "threads" exist, they are just assuming the Third Law in their definitions. And, of course, where action at a distance is concerned, there are no such threads at all.

The laws of acceleration, $F = ma$, and of the composition of forces are, says Poincaré, *conventional.* But they are not arbitrary conventions. For our imperfect experiments justify the adoption of these conventions. Like so much in Poincaré's conventionalism, the reader is left here in some perplexity. The overall line seems to be that we can adopt the standard laws and assign magnitudes to the non-directly measurable quantities (F and m in particular) so as to make the laws come out approximately true for nearly isolated systems. But, the line continues, since we have freedom in assigning the magnitudes of these quantities, we can always save apparent violations of the laws by modifying the values of F and m that we assign. So adopting the laws may be a useful formalism for characterizing the world, but the laws themselves can never be refuted by experiment. We could, in fact, pick other laws and other assignments, but to do so would be to complicate our descriptive–predictive apparatus.

The mere fact that the standard laws of dynamics were soon to be rejected in relativistic (not to speak of quantum) theories surely shows that something is not quite right with Poincaré's position. And the way in which that position has been presented leaves the reader puzzled as to exactly what Poincaré's claims come down to here. Nevertheless, his incisive critique of naïve methods for determining mass and force values "up from the ground of observation" remain of deep importance.

The second essay on dynamics is devoted to the hoary problem of absolute versus relative motion. Start with the fundamental principle that a system will obey the same laws of motion whenever its motions are referred to any reference frames that are in uniform, rectilinear motion with respect to one another. For an isolated system differences of acceleration will depend only upon the relative positions and velocities of the components of the system and not on absolute velocities – at least so long as the motions are referred to an inertial reference frame. Not surprisingly, Poincaré will have us believe that this principle is neither a priori nor experimental, but has the same status as the other laws of motion discussed in the previous essay.

But why does the principle not hold for rotating reference frames? There, plainly, are the well-known effects of "absolute accelerations." Newton would have us believe in absolute space, but must we accept that? Consider a clouded Earth from which we never saw the fixed stars, that visible framework of material objects that serves as a good visible inertial frame. Could a Copernicus ever arise to declare the Earth to be in rotation? Such experiments as that with a Foucault pendulum would cry out for explanation. We could try explaining the motions by positing "real" centrifugal and Coriolis forces. But how complex these would be. The centrifugal force would increase with distance from the Earth's center. And the Coriolis forces would require abandoning left–right asymmetry for the real forces of the world. Surely a Copernicus could come along and posit, instead, a rotating Earth.

But "the Earth rotates," has "no meaning" if rotation with respect to some (non-existent) absolute space is meant. It is much better to say "It is more convenient to suppose that the Earth turns round, because the laws of mechanics are thus expressed in a much simpler language. That does not prevent absolute space – that is to say, the point to which we must refer the Earth to know whether it really does turn round – from having no objective existence . . . these two propositions, 'the Earth turns round,' and 'it is more convenient to suppose that the Earth turns round,' have one and the same meaning."

To illustrate further, Poincaré has astronomers deal with a solar system, again having no visible reference to the fixed stars to pick out a convenient inertial frame. In order to get the right results for their orbits they will need not only the initial positions and velocities of the individual planets, but also attributions to the system as a whole of an appropriate angular momentum with respect to the frame they choose. This is because, without the fixed stars to pick out an inertial frame, they can deal only with relative, as opposed to absolute, longitudes for their planets. If they were aware of only one solar system, they might take this constant to be an "essential" constant of nature. But, on discovering more than one system, they would discern it to be a number that varied from system to system.

When we explore the entire Universe we may need to posit such numbers as well. Were we to find ourselves doing so, we could, if we wished, talk about the absolute orientation of the Universe with respect to space itself, or, more importantly, the rapidity with which this orientation varies. That is, we could talk about the rate of spin of the material cosmos with respect to absolute space. Or we could fantasize some mysterious reference object instead of space itself to do the trick (Neumann's "body Alpha"). This would be ". . . the most convenient for the geometer. But it is not the most satisfactory for the Philosopher, because this orientation does not exist." But so long as our theory carries us correctly from observation to observation, we have nothing to complain of.

Once again Poincaré moves us to thought with his insights. Certainly his claim that one could come up with a correct predictive theory of classical dynamics that never mentioned space itself in its metaphysics, but that did take account of the need to add constants of angular momenta for systems as a whole should one choose to work in an absolutely rotating frame, is a useful step for a metaphysical relationist. Nevertheless, all of his talk of the mere convenience of working in inertial frames will not truly satisfy the Newtonian who is demanding a physical explanation for the preferential simplicity of the inertial frames, and who is finding that explanation in the structure of "space itself." Nor does Poincaré's dogmatism about the non-existence of such a structure, if that is, indeed, what he is denying, seem convincing. Here the advances obtained by later moving to a space-time framework for classical dynamics itself, a matter we will explore in Chapter 19, will be of some help.

In 1922 Paul Painlevé published an interesting critical exploration of the foundations of dynamics. He found a root insight in what he called the "principle of causality." This goes beyond the usual philosophical determinist's assertion that the initial state of a system will fix its future states. It demands also that there be an invariance of evolution over transpositions of the system in space and time. If we start a system in a state at one time and it evolves through other states as time goes on, a similar system at some other initial place and time, started with the appropriate translated initial state, must evolve through a sequence of states that are the appropriate space and time translations of the states the initial system evolved through. Interestingly, Painlevé does not seem to require invariance under change of orientation as well in his basic principles.

Painlevé takes it that both "the Copernicans" (among whom he includes Copernicus, Kepler, Galileo and Newton) and "the Scholastics" (the Aristotelians) shared this principle. They also, he believes, shared the conviction that particles infinitely removed in distance from a test particle could have no influence on its motion. And, he thinks, they shared the idea that this principle of invariance applied only to the absolute states of systems, not their mere relative states with respect to other systems.

Where the two schools differed, he claims, is in what they took to be the absolute dynamical states of systems. For the Aristotelians this was just absolute position. The influence of nearby objects was to change the absolute position of the test system by giving it some motion. For the Copernicans the absolute quantities were absolute position and absolute velocity. The influence of the nearby objects was to change the absolute velocity (to induce an absolute acceleration) on the test system.

To make sense of these notions we need ways of determining the proper scheme for measuring distances and time intervals, and we need to refer our positions and motions to reference frames. How can this be done when the "absolutes" that we need are not available to us, only measurements of space using material rods, of time using clocks and of motion relative to material reference frames? The key to classical dynamics is found in the realization that there are the following fundamental empirical facts: There is a way of measuring distance, a way of determining time interval, and a way of selecting reference frames such that, when the measuring and referring are done in this "proper" way, the principle of causality becomes true at all places and times for systems characterized in the appropriate absolute dynamical terms.

The Copernicans realized that the appropriate reference frame could be found in one that had its origin near the center of the Sun (the center of mass of the solar system actually) and whose orientation remained fixed relative to the fixed stars. And they also realized that any other reference frame that moved in a straight line and with constant speed relative to this one would serve equally well as the reference frame relative to which positions and velocities could be considered "absolute."

One could then offer "positive" axioms for dynamics by making this implicit structure of "Copernican" thought explicit. Take as basic the claim that, for all of the particles of the Universe, their initial absolute positions and velocities determine their future absolute values. Next take as basic the causality principle that any dynamical motion remains invariant under a translation in space and in time. Third, assume that the motion of any particle is unaffected by anything related to the position or velocity of infinitely distant points. It follows from these axioms that any "isolated" system follows a straight-line path. The Copernicans "complete" this corollary with the assertion that the speed along the path is constant.

So read dynamics as first saying the following: There is a way of measuring spatial distance, and a way of measuring time interval, and a frame of reference relative to which positions and motions are referred, such that a material particle infinitely removed from all others follows a straight-line path at constant speed.

Next add the action–reaction principle in the following guise: The acceleration induced on M_2 by M_1 is opposite in direction to that induced on M_1 by M_2. These accelerations are determined in magnitude and direction by the spatial separation and relative velocities of M_1 and M_2. The ratio of these accelerations is independent of the positions and velocities of the masses. And, finally, the effect on M_1 of a number of particles is the geometric (vector) sum of the actions of each independently.

That the fixed stars form an appropriate reference frame, and that various "good" measuring rods and clocks "correctly" measure spatial and temporal intervals, i.e. that referring motions to such a frame and measuring intervals with such clocks results in stipulations of positions, velocities and accelerations such that the inertial and action–reaction axioms turn out to be true, constitutes the empirical content of our dynamical science. Of crucial importance empirically is also the fact that any other reference frame in uniform rectilinear motion with respect to that standard frame would also be one in which the axioms held.

Painlevé offers us clear insights into the fundamental role played by invariance principles in the foundations of dynamics. His suggestion that insights into such invariance were implicitly present in the minds even of the originators of the theory is a fascinating one. And it is certainly true that invariance under what is later called the Poincaré group, namely the combined actions of spatial translation and rotation, of temporal translation, and of "boosts" from one inertial frame to another, is key to a formal reconstruction of the theory at its foundations.

SUGGESTED READING

An outline of the thoughts of Helmholtz, Kirchhoff, Saint-Venant, Reech and Andrade is given in Part IV, Chapter 10, sections 1–4 of Dugas (1988). Hertz's "neo-Cartesianism" is fully explained in Hertz (1956) and outlined in Part IV, Chapter 10, section 6 of Dugas (1988). For Saunders' critique of Hertz see Saunders (1998). A collection of pieces dealing primarily with Hertz as a philosopher of science and his approach to theories in general is given by Baird (1998). Mach's approach to mechanics is treated at length in Mach (1960). For his general philosophy of science, see Mach (1959). An extended treatment of Mach from an historical perspective is given by Blackmore (1972). Blackmore (1992) contains a wealth of original documents from Mach and commentators on him. Bradley (1971) focuses on Mach's general philosophy of science. Poincaré's insights into mechanics can be found in Chapters VI and VII of Poincaré (1952). Painlevé's ideas are summarized in Part IV, Chapter 10, Section 9 of Dugas (1988).

CHAPTER 17

New directions in the applications of dynamics

Throughout this book so far our attention has been firmly directed to issues arising out of consideration of the fundamental postulates of dynamical theory. What should those posits be? How are the alternative choices for foundational axioms related to one another? What are the basic concepts utilized by those posits and how are they to be interpreted? What is the epistemic status of the basic postulates? And, finally, what kind of a world, metaphysically speaking, do the posits demand?

In this chapter, however, we will look not at the fundamental posits but, instead, at how the consequences of these posits are developed in particular applications of dynamical theory. We have generally been avoiding issues of application. But some of these realms of application are quite fascinating from an historical point of view, for it is often an extremely difficult task for the scientific community to figure out how the elegant basic posits of the theory are actually to be used in describing the complex systems we find in nature. Great efforts in classical dynamics were devoted to such issues as the appropriate frameworks for applying the theory to the motion of rigid bodies, to fluids of ever more complex nature, going from the non-viscous to the viscous and from the incompressible to the compressible, and to more complex continuous media. Difficult special cases, such as those of shock waves, where familiar continuity assumptions cannot be maintained, also led to the development of subtle and sophisticated application programs for the dynamical theory.

To deal with any of these issues in a way adequate to their historical complexity would plainly be beyond the scope of this project. But a few programs for the application of the theory do merit our attention here. In this chapter we will be looking briefly at three such programs: the development of qualitative dynamics out of the program of celestial mechanics, the growth of chaos theory out of qualitative dynamics, and the blending of probability into dynamical theory in statistical mechanics.

The reason for focussing on these three areas of application is that in each case the scientific program of trying to apply the fundamental theory to novel areas resulted in the realization that new aims for explanation were required rather than those that had dominated in the earlier cases of application. These new explanatory goals called out for structures of explanation unlike those familiar in the earlier cases. Since one of the aims of this book is to explore how the changing scientific theory brought along with it changing notions of what a scientific explanation ought to be, these particular cases of the exploration of the application of the theory merit our attention.

We will not, however, explore these three areas to the depth that they deserve. This is for two reasons. The first is that a serious historical and methodological investigation into these problem areas would require the production of individual works each potentially as large as this whole book. The second is that many of these issues have been explored at some length in the existing literature, and here there is no need to repeat the work that has already been done. After some preliminaries, then, we will proceed to look at each of these areas of application of dynamics only in a cursory way, pursuing the science only far enough to gain some broad insights into how the changing science brought with it changing modes of the notion of what an explanation in theoretical science might be like.

17.1 FROM CELESTIAL MECHANICS TO QUALITATIVE DYNAMICS

What is the form of a typical question to be posed to dynamical theory? And what is the typical form of an answer to that question? The dominant idea for centuries was, of course, the view that the place of dynamics was to allow us to predict future positions and velocities of objects from present positions and velocities.

The fundamental equations of dynamics are differential equations. In the Newton–Euler format these relate accelerations (linear or angular) to forces and torques. Of course, one needs to supplement the fundamental dynamical principles with those that describe how forces and torques are generated. These additional principles are taken to relate the forces and torques to the relative positions and velocities of the components of the system. Given both of these sets of equations, a specification of the initial positions and velocities of the components of the system will "generate" their future positions and velocities. Or, rather, the fundamental mathematics of differential equations will guarantee that these future values are determined for at least a "short" time after the initial time chosen.

But it is one thing to know that such a solution to the differential equation exists and quite another to be able to find such solutions. In fact, it became apparent quite early on that the number of problems for which exact solutions to the differential dynamical equations could be found, solutions that could be expressed in terms of such familiar functions as closed polynomials, exponentials and trigonometric functions, was very small indeed. In point-particle mechanics, where systems are idealized as being made up of interacting point masses, one could solve such problems as that of two bodies interacting by their mutual gravitational attraction and the motion of a particle in a central force field or in a force field with two centers of attraction. In rigid-body dynamics one could solve the problem of a general rigid body in the absence of external forces or of the axially symmetric top in a gravitational field. In fluid dynamics and in more complicated problems of continuum mechanics one could give exact solutions when very stringent symmetry conditions held for the systems in question.

But exact solutions were few and far between. Even for the problem of three bodies interacting by means of their gravitational attraction, closed exact solutions seemed out of the question. One could try to deal with such problems instead by various methods of approximation, and these were soon devised. Expanding solutions in terms of appropriate power series, and trying to derive new solutions as modifications or "perturbations" of known exact solutions became commonplace. But even here life became very difficult indeed.

Newton, for example, tried to understand how the influence of the Sun perturbed the otherwise elliptical orbit of the Moon around the Earth. The Moon's line of the apsides, the line connecting its nearest and furthest points from the Earth, rotates relative to the fixed stars. Newton's attack on the Moon problem found such a rotation – but only half of the observed value. Newton remarked to a contemporaneous astronomer that the problem of the Moon was the only one he had worked on that had made his head ache.

The solar system was the main object of investigation where issues of how to treat problems for which exact solutions were not forthcoming arose. Newton himself realized, of course, that the Keplerian orbits for the planets could not be exact. Indeed, he takes the perturbation of the planets' motions by their mutual gravitational interactions as one important piece of evidence in the chain leading to the hypothesis of universal gravitation.

The solar system provides also the first example of how these difficulties in finding exact solutions lead to the suggestion that it is not the exact future development of a system that we should be searching for, but, rather, the answer to a "qualitative" question instead. Suppose it is a very difficult matter to predict exact positions and velocities of the planets into the distant future. Could we not at least expect to answer the question as to whether or not the system is stable? How could it be "unstable"? Well, the perturbations by the planets of one another's orbits might throw some planet into collision with the Sun or with another planet. Or some planet might attain sufficient energy of motion as to be thrown out of the Sun's influence altogether, that is, the planet might leave the solar system altogether. Can dynamics reassure us that these untoward events won't happen?

The problem of the stability of the solar system became a classic question for the very best theoretical physicists of the eighteenth century. A standard approach to the question developed. As the planets circle the Sun they tug on each other. These tugs will obviously result in periodic changes in the planets' distances from the Sun. But will there be "secular," i.e. non-periodic, changes in such parameters as the mean distance of a planet from the Sun as well? If that mean distance decreased or increased non-periodically over time, the stability of the solar system would plainly be in question.

This problem was taken up by a succession of researchers using ever subtler power-series methods. The central idea is to assume that the secular changes, if any, will be "slow" relative to the "fast" changing periodic pushes and pulls the planets exert on each other as they move in their orbits and the "fast" changing periodic changes in the planetary parameters these tugs induce. In 1773 Laplace found no secular terms in a first-order series approximation. Shortly thereafter Lagrange found no secular terms arising for all-order approximations using expansions in the eccentricities of the ellipses and in the inclinations of the orbits of the planets to one another and in the first order of expansion using the ratio of planetary mass to that of the Sun. In 1809 Denis Poisson improved on this by finding no secular terms in the second-order terms of expansion with respect to mass. But in 1878 Spiru Haret found that in the third-order terms with respect to mass secular terms in the values of the major axes of the planetary orbits did appear.

In any case, the problem remained a great puzzle. The results for the first few terms of an expansion couldn't tell you what the full series would entail. Indeed, there were grave questions about whether, in fact, the series approximations used converged, that is, whether they gave finite predictions even when summed to all orders.

The problem here is that of "small divisors." When the terms in one of these series are calculated they have ever decreasing numerators the further one goes out in the series. But the denominators of the terms contain factors of the form $jn_1 + kn_2$, where j and k are integers and n_1 and n_2 are the periods of the planets. These terms can be very close to zero when jn_1 and kn_2 have similar absolute values but are opposite in sign. The physical reason for this is clear. If the planets are such that after j orbits of one and k of the other they come back to a repeated situation of closeness in their orbits, the force exerted by one planet on the other can continually add up and become significant, even if each individual "pull" or "push" is small. This is the situation of resonance, which will be familiar to anyone who has imparted large motions to a child on a swing by repeated small, but well-timed, pushes. It is impossible to tell by any kind of superficial inspection whether the numerators in the series will become small sufficiently fast to overcome the resonances possible for large values of j and k.

The early researchers also realized that the general problem of n bodies was far too complicated for them to deal with. First steps would have to be taken with the simplest case beyond the two-body problem – the three-body problem. Even simpler was the restricted three-body problem. Here one mass is considered much smaller than the other two. The two larger masses are taken as moving according to the solution of their two-body problem, and the third as moving under their influence without influencing them. If one goes further and takes all the masses as moving in a single plane, then one has the even simpler planar restricted three-body problem. But even these simplest cases proved rich with complexity.

Euler and Lagrange found particular solutions to the three-body problem in which the three masses remained collinear and in which they remained at the vertices of an equilateral triangle with respect to one another. For a very long time these "central solutions" were the only exact solutions known to a three-body problem.

Despite these profound and fundamental difficulties, the great explorers of celestial mechanics from Laplace and Lagrange, through Poisson and Jacobi, to such nineteenth-century figures as C. Delauney, Hugo Gyldén, Anders Linstedt and G. W. Hill were able to make significant gains in developing series approximation methods combined with brilliant novel ways of representing the dynamical problems that allowed them to explain many of the deviations of the solar system from ideal Keplerian orbits. They were able, for example, to calculate the not insignificant perturbations the large planets (Jupiter and Saturn) exert on each other and were able to make progress on the very complicated problem of the motion of the Moon. Another famous success of the method was the prediction of the existence of Neptune, and of its position, as a perturber of the orbit of Uranus.

The great turning point from celestial mechanics to qualitative dynamics was the work of Poincaré. Beginning with early papers on two-dimensional dynamics, continuing through his epoch-making prize essay on the three-body problem, and going further with his work in his *New Methods of Celestial Mechanics* and his *Lectures on Celestial Mechanics*, Poincaré set the course of the application of dynamics to many-body problems in brand new directions.

Poincaré's first and most basic idea is to look not at individual trajectories but at whole classes of them. Frame a problem in terms of a number of point masses and their interaction. Then consider what evolves from all possible initial positions and velocities for the masses at once. Do this by looking at the trajectories that go through each initial point when the system as a whole has its state represented by a single point in the multi-dimensional phase space for the entire system. Seek to discover not the details of each orbit, but the structural features of the "phase portrait" as a whole. Here such features as the number and distribution of periodic trajectories, and, especially, the number, structure and distributions of singular points in the phase flows, become crucial. For it is these features that govern such basic issues as whether or not collisions take place and whether or not long-term motions can be classified as stable or unstable.

A major problem whose beginnings were explored by Poincaré is that of structural stability. Suppose one has the phase portrait of a system. Now change slightly some parameter of the system. Vary slightly, say, the mass of one of the point particles. Will the phase portrait remain the same (qualitatively) or will there be radical changes, say the appearance or disappearance of some of the singularities in the flow? When a small change in parameters gives rise to a structural change in the phase portrait, how can we grasp in our theory the occurrence and the nature of such structural changes? These changes are often called "bifurcations."

Poincaré avails himself of both old and new mathematical methods in his explorations. He continues the use of the methods of analysis, the methods developed to solve the differential equations of the theory. In particular he is a master of the game, familiar to celestial mechanics, of expressing solutions in terms of power series and then changing variables and rearranging terms to make these series tractable means of solving the problem at hand. But he also avails himself of more abstract methods. He utilizes results depending only upon ways of measuring "volumes" of regions in phase space, and how these volumes change as one follows the points in them under their dynamically induced motions. In other words he initiates, with earlier anticipations by Liouville of which Poincaré was not aware, the so called "measure-theoretic" approach to dynamics. And he avails himself of methods that rely only upon the mathematics of spaces that depends upon the continuity of regions and the preservation, or non-preservation, of that continuity by mappings of the space onto itself. That is to say, he uses what became called "topological" methods to resolve dynamical problems. Indeed, Poincaré is one of the great founders of the very discipline of topology.

As his predecessors did in celestial mechanics before him, Poincaré focusses his attention on the simplest systems: three-body, restricted three-body, planar restricted three-body and – even simpler – planar restricted three-body problems where the two "massive" bodies are in circular orbits. These systems are simple enough that some progress can be made on them, but, amazingly, intricate enough that their study reveals layer after layer of unsuspected complexity in many-body dynamics.

One of Poincaré's greatest innovations was his invention of what became known as "Poincaré sections." In an n-dimensional phase space, pick an $(n - 1)$-dimensional subspace. Choose subspaces that are nowhere tangent to the orbits of the systems, so that the systems intersect these subspaces "transversally." As the systems "flow" along their trajectories, mark the points where they intersect the subspace over and over again. Now look at the mapping of the subspace into itself that takes a point in the subspace where an orbit crosses it into the point of the subspace where the orbit next intersects the subspace. As Poincaré shows, a vast amount of information about the qualitative dynamics of the system is contained in the nature of this "shift map."

Poincaré also emphasizes the fundamental importance of periodic trajectories. These will generally be a "small" subset of all trajectories. But their number, their nature and their distribution among the set of all trajectories are of fundamental importance. Much structural information lies in how these periodic orbits intersect the Poincaré sections. And the major route one has into finding the solutions for non-periodic trajectories is to look at how these arise as perturbations out of the periodic orbits. One can gain deep insights without even knowing what the exact periodic orbits are, if one can discover results about how many periodic orbits there are and how these are distributed, densely perhaps, among the general class of trajectories. A major result of Poincaré was to find a condition on the shift map on a section, framed in topological and measure-theoretic terms, such that if it were satisfied one could prove the existence of an infinite set of periodic trajectories for the restricted three-body problem. This work was brought to a culmination later by George Birkhoff who proved that the needed condition held.

One of Poincaré's early results was negative. Those working on the three-body problem often hoped that some additional constant of motion could be found beyond those known, and that one could fix this constant by means of a convergent series approximation in terms of some parameter of the system (say the small mass of the third body). Poincaré showed that at least the kind of additional constant that was usually looked for could not exist. This suggests the possibility of wildly uncontrolled motions even in the three-body case, but the real demonstration of that came later.

Another early result of Poincaré was to use measure theory together with the dynamics to prove the existence of the integral invariants we mentioned earlier in the chapter on the Hamiltonian formalism. In an extremely elegant and simple proof he showed that any dynamical system where the energy was conserved, and that met the constraint that the region of phase space entered by the trajectories was bounded, would of necessity meet a weak stability condition. This means that for all but possibly a set of initial points of measure zero, given a region in phase space about that point however small, the trajectory from that point would reenter that region an infinite number of times. The condition in question applied to the three-body problem, showing that a kind of "almost recurrence" would be inevitable for almost every initial condition of such a system. The existence of these integral invariants (which, as has been noted, was anticipated by Liouville) became essential for later statistical mechanics.

Poincaré was also a master of the game of manipulating the infinite series that appeared in traditional perturbation theory. But, not surprisingly, his main concerns were not with the use of particular series to solve particular problems, but, instead, with general facts about the kinds and natures of the theories that arose in celestial mechanics. Terms in series that had the time parameter appearing outside of a periodic, trigonometric function were disturbing. For they threatened to "blow" up terms in the series as the time became longer. Poincaré was able to show that many such terms could be eliminated by clever choices of variables and expansions. But, he warned, the absence of such terms was no guarantee that the series would converge as the number of terms in the series became ever larger. Even if all of the individual terms were periodic in nature, the sum of the terms could still diverge.

Poincaré pointed out that "convergence" meant quite different things to the "practical" astronomer and to the pure theoretician. The former wanted a series that was such that a very few initial terms would provide excellent approximations to the actual motion of perturbed bodies for limited times. The latter wanted series that would converge even after an infinite number of terms had been summed and that would be convergent for all times, for the theoretician was interested in "stability in the long run." In studying these matters Poincaré made important advances in our understanding of how infinite series could be used to solve differential equations. A series could be convergent, but not uniformly convergent. Uniform convergence demands, crudely, that the rate at which the terms in the series decrease be independent of the value of the time parameter in them. It is usually uniform convergence that theoretical qualitative dynamics demands.

But the practical astronomer need not even demand convergence. Suppose that a series actually diverges. The series then does not approach a finite limit as the number of terms becomes infinite. But suppose that even a few terms will give results that become very accurate as the time parameter becomes very large. Such a series, later called "asymptotic," will suit the practical astronomer's purposes just fine.

What was, perhaps, Poincaré's greatest discovery had a truly remarkable history. His submission for a prize competitition organized by Magnus Mittag-Leffler won the award of one thousand crowns and was duly printed. Before it was released, however, Lars Phragmén, a reviewer, suggested to Poincaré that one of the proofs was problematic. While reflecting on this, Poincaré found a deep flaw in a fundamental result. In correcting this, he immediately realized that he had discovered a radically new and vitally important piece of qualitative dynamics, nothing less, actually, than the beginnings of what is now called chaos theory.

In studying two-dimensional Hamiltonian flows Poincaré showed that only two kinds of singularities could exist. In one of these, orbits flowed around a fixed point. In the other, a fixed point was a "saddle point." It had a line through it in which flows approached the point and a different line where flow was away from the point, and all other nearby orbits swept in, neared the point, and then swept away again. (Vector fields in a plane can have other kinds of singularities, but not when they are described by a Hamiltonian equation.)

Suppose we have an orbit that is periodic and that intersects its section in a single point. Suppose there is a surface in which orbits asymptotically spiral in toward the periodic orbit, and a distinct surface where they asymptotically spiral away from the periodic orbit. These surfaces will intersect the section in lines through the point of intersection of the periodic trajectory, and the shift map of the section as systems go through it time after time will have a saddle-point singularity.

Poincaré initially though that if one extended the lines in the plane (the sections of the so-called stable and unstable manifolds of the periodic trajectory) they would join up, giving rise to the possibility of some stability proofs. His second thought was to realize that these lines could intersect transversally. If they did this once, he realized, they would have to intersect an infinite number of times. Worse yet, the "loops" they made between intersections would have to become narrower and narrower, but longer and longer, as the points of intersection asymptotically approached the point of intersection of the periodic trajectory. These intersection points he called "homoclinic points."

What Poincaré realized is that the web made by these manifolds, the "homoclinic tangle," indicated an enormously complicated set of trajectories near to the original periodic orbit. His often-quoted observation on this is worth repeating here once more:

> If one seeks to visualize the pattern formed by these curves and their infinite number of intersections, each corresponding to a doubly asymptotic solution, these intersections form a kind of lattice-work, a weave, a chain–link network of infinitely fine mesh; each of the two curves can never cross itself, but it must fold back on itself in a very complicated way so as to recross all the chain-links an infinite number of times.
>
> One will be struck by the complexity of this figure, which I am not even attempting to draw. Nothing can give us a better idea of the intricacy of the three-body problem.

From Poincaré onward, then, celestial mechanics becomes a new discipline, namely qualitative dynamics. New mathematics, in the forms of measure theory and topology, supplements the older methods of analysis. New devices, the Poincaré section and shifts on it, for example, become the stable method of the subject. The questions asked are no longer so much how to find an appropriate approximation method for determining the details of some celestial orbit, but, rather, the qualitative, often global, questions about the nature of the phase portrait for some especially interesting, or especially tractable, dynamical structures. We will look, once again far too briefly, at how two such subject areas, the study of singularities and the study of stability, developed after Poincaré and then at how two whole disciplines, the theory of chaotic systems and the theory of statistical mechanics, arose out of qualitative dynamics. And we shall explore, again far too briefly, how these disciplines brought with them new types of "Why?" and "How?" questions and new types of answers to them into dynamics.

17.1.1 Singularities

Start a dynamical system in an initial state. The basic nature of the differential equations of dynamics guarantees that they will predict a new state for the system over a "small" future time interval. But they do not entail that a solution to the equations can be projected from the initial data over any arbitrarily long period of time. If the solutions cannot be projected for all future time, the solution from those initial conditions is said to possess a singularity.

For systems composed of point particles, the systems of the n-body problems, one obvious possibility for a singularity comes from the possibility of two or more of the particles colliding. Or, if the notion of *point* particles colliding bothers one, one can consider this as the possibility of the limit of the distance between the particles going to zero in some finite time.

There are several kinds of questions qualitative dynamics is going to ask about collisions. Can a system of a given kind ever suffer a collision? If collisions are possible, will they arise only from certain initial conditions? If so, what does the class of initial conditions that result in collisions look like? What will the collisions be like? Will there only be pairs of particles colliding or are multiple-particle collisions possible? Will collisions result in other important consequences, for example, will a two-particle collision in a three-body problem result in some special motion of the third particle? In general, as the particles near collision, what, qualitatively, does the space of trajectories headed for collision look like? If the particles do collide, is there some "natural" way of extending the predictions of dynamics past the collision? All of these questions have been explored at length, especially in the case of the three-body problem.

Even if there are no collisions, is there some other conceivable way in which the extension of the solution past some finite time may become impossible? Can singularities of such a non-collisional nature actually occur for some kind of dynamical system or other? Can one actually construct a model dynamical system that shows such a non-collisional singularity? Here again much important work has been done and fascinating results have been found.

For the three-body problem Paul Painlevé was able to show that the only possible singularity was a collision, and that if no collision occurred the dynamical problem was then solvable by using a convergent power series. He made conjectures about what the class of initial conditions leading to a collision looked like, but was unable to prove them. Tullio Levi-Civita was able to prove the existence of the appropriate characterization of these initial conditions for the restricted three-body problem, a result extended (with some assumptions) to the full three-body problem by Giulio Bisconcini.

The high point of this work was the astonishing result obtained by Karl Sundman. He showed that, for three bodies, no triple collision was possible unless the angular momentum of each body was zero. Neglecting that special case, he could also show that, if one took the double collisions as being between two bodies that elastically rebounded after the collision, one could extend the solution of the equation through the collision by what is know as analytic continuation. He was also able to show that one could not have an infinite series of double collisions that had a finite limit point in time. This allowed him to take the solution as analytically extendable through the individual collisions without limits on the number of collisions. He was then able to make a series of transformations of variables in a series expansion to the general three-body problem that resulted in a series that he could prove always converged for all time!

Is the three-body problem then solvable? If you are asking whether one can find a solution to the equations in terms of familiar functions, the answer is "No." If you are asking whether one can find a perturbation series valid for all times, even when there are collisions (assuming analytic extension through them), then the answer is "Yes." Is that series solution useful for astronomical predictions? Now the answer is "No," since the series converges "very slowly." How slowly? Well, one estimate has it that it would take the calculation of $10^{8000000}$ terms to provide any information of value to the practical astronomer! If your question is whether it is possible to provide a complete characterization of the phase portrait for all possible initial conditions, the answer is, once again, "No."

Modern computer studies of collisions in three-body systems show one or more particles often being thrown out to great distances at high velocities after two particles have approached their collision. As early as the 1920s, though, Jean Chazy made deep investigations into what kinds of motion would follow a collision in a three-body system. Putting the bounded and oscillatory motions to the side, he was able to offer elegant qualitative characterizations of the kinds of situation in which either one or two of the mutual distances between the particles increase without limit.

For a three-body system, neglecting the exceptional case, only two-particle collisions are possible. But for general *n*-body systems we can imagine collisions involving three or more particles. What would these be like? A great deal of insight has been obtained about three-particle collisions.

Sundman was able to show that in a three-body collision the particles asymptotically reached a central configuration prior to the collision. In such configurations the particles remain in a fixed pattern that varies only in its scale, ever shrinking as the particles approach collision. Aurel Wintner showed that this result held for collisions of more than three particles. Carl Siegel studied three-body collisions and showed that the set of orbits that reached such collisions formed a smooth sub-manifold of the phase space. He also was able to show that for almost all masses assigned to the particles it was not possible to extend solutions through such collisions by the means of the method of analytic continuation that worked for two-body collisions.

A new approach to the study of multiple-particle collisions arose out of a suggestion of Robert Easton. Instead of asking whether solutions could be continued through collisions by looking for power-series expansions that would join together solutions before and after collisions (the method of analytic continuation of solutions through collisions), ask a different question. Look at the trajectories that systems follow that come close to collisions but just avoid them. Ask whether there is a unique and continuous way of embedding a collision solution into that set of non-collision orbits.

For the two-particle collisions, Easton's method provides the same solution as that using analytic continuation. The trajectories of two particles near collision become ever more elongated elliptical orbits, in which the two particles move in opposite directions with approximate conservation of linear momentum along a line. The trajectory of collision that has the two particles elastically rebound from one another obeying the usual collision law of conservation of momentum, the trajectory corresponding to analytic continuation, is then the trajectory that fits continuously into the pattern of near-to-collision trajectories.

Other simple examples, however, show that the two methods will generally not result in the possibility of finding a unique way of projecting trajectories beyond collisions. What happens when three particles collide? Here there is, in general, no solution by analytic continuation. What results if one looks at triple collisions from the Easton perspective? Deep results on this problem were obtained by Richard McGhee. He used a method that sought the description of orbits approaching a triple collision by a process of continually rescaling space and time variables as the orbits got closer and closer to a collision. The result is the production of a "collision manifold," a structure that characterizes the set of orbits that come near to a collision. The conclusion that follows is that, for almost all assignments of masses to the particles, triple collisions are not "regularizable" in Easton's sense. One cannot find "nice" ways of embedding a unique collision trajectory continuously into the set of near-collision orbits.

One way of gaining insight into this result is to think about the problem of one body moving down a line and approaching two other bodies that are near each other. If there are elastic collisions the system will end up with the first body near the one in the middle and the one on the far end flung out. But then even a slight change of initial conditions (exchanging the near rest ball for the far one) will result in a radically changed final outcome – a different mass will end up flung out from the remaining pair. Something like that happens when the masses are interacting by gravity as well. A typical result of the near-collision orbits has two particles going into close orbits with the third being flung out to a great distance. But, for almost all masses, which particle will leave the vicinity of the other two will sensitively depend upon the exact initial conditions of the particles. Under these circumstances it will be impossible to find a unique trajectory for continuing orbits past a collision that will fit nicely into a well-structured neighborhood of near-collision orbits.

For the three-body system collisions are the only possible singularity for the system. But what about the situation where there are four or more particles? Here a result of Hugo von Zeipel becomes crucial: If there aren't collisions, the only other possible singularity occurs when one of the mutual distances between the particles increases without a bound within a finite time. In other words, the only non-collision singularity is when one or more particles get "flung out to infinity" in a finite time. But can this ever actually happen? Solutions were found to a system of four particles bouncing off one another along a line that sent a particle to infinity in a finite time, but could you ever have a non-collisional singularity in a system without collisions?

In the early 1990s Zhihong Xia found a marvelous solution to a five-body problem. Four of the particles move in two pairs of parallel planar binary orbits. The fifth oscillates through these orbits at right angles to the planes. Each time number five goes through a pair, it pulls them in, shrinking their orbits and picking up kinetic energy. It goes out further than before and then comes back, doing it all so fast that, despite its longer swings, it comes back faster than the last time. Then it goes to the other pair and does the same thing. Before a finite time limit has been reached, the particle has oscillations of any length whatever, and the solution cannot be extended beyond the limit time. The proof is immensely subtle, requiring that one subtract from initial states all those that result in the particles, for example, colliding, and showing that the ones left that result in the non-collision singular solution form a set of non-zero measure. Additional solutions by J. Gerver, in which a particle is flung from angle to angle of a regular polygon in a plane, at each angle of which there is a pair of masses in a binary orbit, with the main particle picking up energy from each orbiting pair, the polygon getting larger and the main particle moving ever faster, also are finite-time, non-collision singular solutions.

The study of singularities in qualitative dynamics opens up one area of research that reveals how much rich structure and how many novel explanatory possibilities are latent in applying the fundamental dynamical equations even to extraordinarily simple dynamical systems. Studies of stability reveal additional realms of behavior and its explanation latent in the fundamental dynamical laws.

17.1.2 Stability

We have noted how, in its very origins, qualitative dynamics arose out of questions about the stability of the solar system. Not surprisingly, then, issues of stability always remained a vital topic in this new field of the application of dynamics. As it turns out, the issue of stability is a complex one indeed. There were, in fact, many different notions of stability waiting to be distinguished, and many new and surprising methods and results to be uncovered.

We have already noted a number of contributions Poincaré made to the investigation of stability. There was his discovery of "recurrence," what he called Poisson stability, where a kind of stability followed from abstract measure-theoretic considerations alone. There was his vital discovery of the occurrence of radical instability even in very simple dynamical systems, a discovery that arose out of his correction to his prize paper. In addition, as we have noted, there was his important work on series expansions in perturbation problems, where issues of convergence were explored. In particular his investigations into the problems of resonance and small divisors were seminal.

Poincaré also made progress by investigating the nature of trajectories close to periodic trajectories using his ingenious method of what are now called Poincaré sections. Look at how the intersections of orbits with a transverse subspace change from orbit to orbit. Consider the mapping of these points from one intersection time to the next. Then, assuming that this transformation is sufficiently regular, approximate that mapping by a linear transformation. It will then turn out that the study of how the orbits behave near to the periodic orbit (which generates a singular point in the transformation since its intersection point always maps to itself) will depend on what are called the characteristic values (eigenvalues) of that linear transformation. Here Poincaré was able to avail himself of earlier work he had done on the possible singularities of solutions to differential equations in a two-dimensional plane. These characteristic values, which are complex numbers, were called by him "characteristic exponents." They characterized, at least locally, how orbits near to the periodic trajectory approached it in time, diverged from it in time, or, in rare cases, moved steadily around it.

Partly inspired by early papers of Poincaré, the mathematician Aleksandr Lyapunov made great strides in developing a systematic study of the stability of dynamical orbits. Lyapunov is generally concerned with a notion of stability far more demanding than Poincaré's notion of recurrence. A solution to a differential equation is stable in Lyapunov's sense if a solution that starts "nearby" to a given solution at a particular time remains near to that given solution for all future times. He also has a notion of asymptotic stability that demands that the perturbed solution be stable but also actually approach the given solution in the infinite-time limit.

Lyapunov's method is generally to look at the difference function between the given solution and the perturbed solution, obtaining a new differential equation for that difference. The behavior of the solutions to this new equation near equilibrium points, solutions that are time-invariant, reveals the information about the stability of the solutions of the original equation.

Lyapunov developed some very general methods for testing for the stability of the solutions of a differential equation. Unfortunately, it is usually very difficult or impossible to actually apply the methods to discover the presence or absence of stability. Certain special cases, however, are very tractable. If the differential equation has constant coefficients, for example, then the stability of the solutions is determinable by looking at a linear transformation and, in a manner that is reminiscent of Poincaré's work but more general, finding the eigenvalues of that linear transformation. The method could be extended in a subtle way to deal with equations whose coefficients were periodic rather than constant.

The stability problem for linear differential equations can be dealt with in completeness. But what about the vast majority of non-linear equations? As far as local issues of stability are concerned, it turns out that for the most part these can be dealt with by assuming at any time a linear approximation to their solutions. So if we are interested only in how stable the solutions are for short time intervals around a given time, the methods applied to linear equations will do the trick even in the non-linear cases. That this is so was shown formally in the 1960s by D. Grobman and P. Hartman. The trick works when the perturbed solutions approach or move away from the given solution, failing only when the perturbed solution circles the given solution without approaching it or moving away from it. That condition is very sensitive to details of the solutions at the time in question, and looking at the linear approximation won't work.

But the wonderful work on local stability, its determination and the characterization of the kinds of stability and instability possible, fails to help with the old global problem of the stability of the solar system. Here new methods are required and these have been found. The new methods are very direct continuations of the methods first employed in the eighteenth century.

Here is the typical problem: Jupiter and an asteroid are circling the Sun in their orbits. We first imagine them having no mutual gravitational interaction. They then both travel elliptical orbits with the center of mass of the system as a focus. Now we imagine the gravitational force between planet and asteroid "turned on." What happens? The orbits certainly must suffer some change. But of what sort? Do planet and asteroid continue on orbits somewhat distorted from their original ellipses, perhaps having, say, slow motions of their perihelions relative to the fixed stars? Or, instead, is the asteroid flung out of the solar system entirely, or is it sent into an orbit that leads to collision with the Sun?

To make progress on this problem required a return to the older methods of dealing with celestial mechanics, namely the use of perturbation theory to describe the influence the heavenly bodies might exert on one another. Early work with these methods seemed to show, in certain special, simple cases, that small enough perturbations would not drive the system into instability, but would, instead, merely result in periodic slight changes in orbits. But more rigorous work was plainly needed. The series used were not shown to be convergent, and, indeed, usually were not. The fact that terms in the series appeared with possible small divisors led to the possibility that even in a series without secular terms individual terms might "blow up" and lead to dramatic instabilities. And the discovery by Poincaré of the radical instabilities in trajectories near to unstable periodic trajectories added to the evidence that even small perturbations might lead to unstable orbits.

The definitive results concerning the fact that in many important cases there is the preservation of stability under small perturbations originated in the work of A. Kolmogorov and were brought to completion by V. Arnol'd and J. Moser. These results consist in a number of theorems collectively known as the KAM theorems.

Traditional perturbation theory worked by looking for a solution to the equations of motion once the perturbation had been imposed. The program went like this: First look at the unperturbed orbits that are "conditionally periodic." Such an orbit is like that of the planet and asteroid with no interaction, each taking a periodic orbit but with different periods. Describe the orbits in terms of action and angle variables. In the case of the planet's the former would be such constants of the unperturbed motion as the semi-major axis of the elliptical orbit, the inclination of the planet's orbital plane, the celestial longitude of its "nodes" where its plane of orbit intercepted that of the plane of the Earth's orbit, etc. The latter would be the celestial longitude of the planet that changed as the planet moved in its orbit. In the case of the perturbed motion the former quantities can vary. But assume they will vary slowly compared with the speed with which the latter change. Then assume that one can derive the change of the former by looking at the average rate of their change over an orbit. The idea is that, as Jupiter and the asteroid move in their orbits, the axes, for example, will suffer periodic changes due to their mutual tugs on each other. But the long-term "secular" change of that axis will be a slow result of the averaging out of those tugs.

Then look for a series expansion that will give the solution as to how the averaged slow variables change over time. Try to show that there will then not be any long-term monotonic change of these corresponding to instability. Ingenious methods are constructed whereby one can find equations in the averaged slow variables that invoke the fast variables only in terms that involve powers of the slow perturbation. One then truncates the series after a certain power and solves, giving, one hopes, a solution to the evolution of the slow variable that is accurate to a certain order in the perturbation.

But will the actual value of the quantity of the orbit really differ from the averaged value only by small, periodic amounts? And is it fair to assume that one can neglect the higher-power terms in the series, given that small divisors might blow up some of those terms even if their numerators (involving powers of the perturbation) are rapidly decreasing?

The KAM methods follow work stemming from S. Newcomb and developed by Poincaré. A successive change of coordinates is found that results in a series in which the terms involving the fast variables in the equation for the evolution of the slow variables are multiplied by factors in the perturbation that decrease with great rapidity. One can then show that these terms drop off sufficiently quickly in their numerators that under the right circumstances the possible effect of small divisors, that is of resonances, in the denominators is suppressed and stability is provable.

One can get some understanding of what is shown by looking at some geometry in phase space. In the unperturbed system the action "variables" are constants of motion. Fixing their values specifies a multi-dimensional doughnut-shaped region, a torus, in the phase space. The orbit in phase space of the point that describes how the whole system is moving is a path on one of those tori. If the periods of the bodies are such that their ratio is that of one integer to another the torus is a "resonant" torus. The paths on it are closed curves. Which curve the system follows depends on the initial conditions. On the "non-resonant" tori, where the ratio of periods is not a rational number, the paths are not closed and they cover the torus evenly over time.

It is resonance or near resonance that causes the small-divisor problem. What the KAM results tell us is that for small perturbations, for orbits initially on most tori far enough from low-order resonance, the perturbation will simply distort the initial torus. The perturbed trajectories will then remain close to the initial ones and stability will hold. The resonant tori will, however, "break up," and all trajectories on them, or on tori too close to them, may very well become unstable. But the resonant tori are a sub-family of all the tori in much the same way as the rational numbers are a sub-class of the real numbers. The resonant tori are dense. As close to a non-resonant torus as one looks, one will find resonant tori. But the set of resonant tori has "zero measure" in the set of all tori. "Probabilistically" speaking, the chance of a torus being resonant is nil. The result is that for small perturbations most unperturbed trajectories remain stable.

Nonetheless, "close" to any stable trajectory is one that will be unstable under perturbation! For small perturbations the set of "Kolmogorov," i.e. stable, tori has large measure. But it is not an "open" set. It is, rather, a Cantor-type set like that which remains if you take the middle third out of an interval, the middle thirds out of the remainders, and so on.

In the "gaps" that center on the resonant tori, the trajectories become wildly unstable. For the lowest-dimensional case, they still are bounded, since they are "trapped" by surrounding Kolmogorov tori. But, in higher-dimensional cases, they can wander off to infinity (Arnol'd diffusion).

Naturally all of this is more complicated than it looks here. For example, for the KAM results to hold a "non-degeneracy" condition must be satisfied. Furthermore variants of the theory apply when the perturbation is itself analytic and when it is not (i.e. depending on how the Hamiltonian function varies with the perturbation). But, all in all, the KAM theory has finally brought into focus the true story about the stability of systems under perturbation that are stable when unperturbed, and thus has set into a brilliant explanatory context the problem of the stability of the solar system that exercised celestial mechanics for three hundred years. It is one of the true triumphs of qualitative dynamics.

17.2 CHAOS THEORY

The results of KAM theory gratify the practitioner of celestial mechanics, for they finally justify the assumption that merely adding a very small perturbation to a system that is without the perturbation integrable and conditionally periodic will not engender instability for every orbit. For small enough perturbations the bulk of the perturbed orbits remain stable.

But instability is still known to be rampant in many cases. Poincaré realized that the existence of his homoclinic points immediately forced the existence of a generic instability for orbits surrounding unstable periodic trajectories when their unstable and stable manifolds intersected transversally. Such a result could be found even for such a simple case as an ordinary pendulum whose axis of oscillation was subject to a small perturbing motion. What was needed was a detailed exploration of such chaotic trajectories. When did they arise? What were they like? How could concepts be framed that would allow our mathematical representations in dynamical theory to give us the right categories for describing such behavior and accounting for its presence?

Some results were obtained by Poincaré himself. His work on how periodic trajectories could have their periods doubled as some parameter changed initiated what later became the systematic study of how chaotic motion arose out of non-chaotic motion when the parameters of a system were varied. His conjectures about systems where periodic trajectories were densely found among the set of all trajectories also marked out the discovery of an important aspect of chaotic systems. George Birkhoff's subsequent proof of Poincaré's conjecture here carried things further.

But for many decades the study of the possibility and nature of dynamic chaos remained stagnant. It was only in the 1970s that persistent and productive attention was turned on the issue. Here work was stimulated by computer experiments of Edward Lorenz. In studying the behavior of a set of simple equations he was using to model atmospheric phenomena, he came upon the appearance of chaotic dynamical behavior in a dissipative system, revealing what later were called "strange attractors." With this work massive attention was once more directed toward the issue of dynamical instability.

These systems are studied by what has come to be called chaos theory. This is really not an independent theory, but, instead, an application of dynamics to a special class of interesting cases. The systems studied are those that Poincaré discovered after having noticed the error in his original prize paper. These are systems that have a number of features that are often taken as defining a chaotic system. The points in phase space representing the system are not divisible into pieces whose dynamical evolution is separable. In particular, there is at least one orbit that comes arbitrarily close to every possible phase point. The systems "fold up" in the sense that there is no "going off to infinity" of the motions. In particular, the systems possess a dense set of periodic orbits – a periodic orbit passes arbitrarily close to any point. Finally, the systems have orbits that are "wild" in the sense that arbitrarily close to any starting point of some orbit there is a point whose orbit "moves away quickly" from the specified orbit over time. Here various notions of differing strength play roles. It can be demanded only that at least one nearby point has a diverging orbit, or that all nearby orbits do. Or it can be demanded as well that orbits separate from one another exponentially fast.

Interestingly, the last condition, the one that demands sufficiently irregular behavior for the orbits that predicting the future exact state of a system from less than perfect knowledge of its initial state is impossible, is often derivable from the first two conditions. But it is this condition that leads to terms such as "chaotic" being applied to the systems in question. Notice that nothing here violates the demand that the systems behave in a perfectly deterministic way. Nor is there any violation of continuity in their behavior.

Much has been learned by studying the cases initially found by Poincaré and abstracting from them. Chaos arises when one has an unstable periodic orbit whose intersection with a Poincaré section is a saddle point. What this means is that the orbit is in two subspaces, one containing orbits that asymptotically leave the orbit in future time and one containing those that asymptotically approach it. It is when those subspaces transversally intersect, forming new homoclinic points, that chaos arises. Reflecting on this, one sees that what one needs is regions of phase-space points that under evolution stretch in one direction and shrink in another, but also "fold back" upon themselves. This insight leads to the "horseshoe map," in which a square is shrunk in one direction and stretched in the other, turning it into a strip, and the strip is bent back upon the original square. Here the evolution under the mapping of points in the square that are initially close to one another bears all of the hallmarks of dynamical chaos.

A further level of abstraction has proven vital in understanding chaotic dynamics. It turns out that one can find a structural mapping, a homeomorphism, between the evolution of points in a chaotic system, with the evolution imposed on infinite sequences of numbers (sequences of zeros and ones will do), that under a shift operation just takes the numbers in one position to some other position in the sequence. A "closeness" relation among the sequences can be defined in terms of how many terms they share in positions near the zero place in the sequence. It then becomes easy to show that there exist sequences whose orbits come arbitrarily close to any sequence, that a dense infinity of periodic sequences exists, and that the shift operation will take sequences that are initially close together into sequences that are far apart – all the characteristics of chaotic motion.

One way of understanding chaotic systems is to contrast their microscopic determinism with what appears as randomness on a macroscopic level. If one breaks up the space of points representing dynamical states into regions, and then asks in which region a system will be found at some future times, it becomes clear that, in general, as time goes on, it will be harder and harder to predict, on the basis of which region the system was in at one time, which region it will be in shortly after. Making predictions about where a system will be requires obtaining information about its initial state that must grow exponentially fast in its precision with the time interval over which the prediction is desired.

Two kinds of system have been studied in detail. First, there are conservative systems in which the energy of the system remains constant. Second, there are dissipative systems in which energy is removed from the system. The latter show orbits that approach some "attractor." The set of orbits will gradually approach some orbit whose dimension is lower than that of the original set of phase-space points. For non-chaotic systems this will usually mean the approach of all the orbits to some stationary point (an equilibrium of the system) or to some nice periodic trajectory to which all the orbits are asymptotic. In the case of chaotic systems the attracting orbit may itself be quite chaotic in nature. Such limiting asymptotic orbits are called "strange attractors," or, better, "hyperbolic attractors." Points on these attractors that are close to one another will, as time goes on, become far apart, so that on the attractor itself there will be increasing difficulty of prediction from imperfect knowledge of the initial state as time goes on.

Further, the attractor may have a curious "fractal" nature, showing a kind of order that appears the same on ever more microscopic scales.

In chaotic systems it may be that there is more than one attractor. A dissipative system will be attracted to one or another attractor depending on its initial state. The set of points for which orbits go toward a given attractor is called the "basin" of that attractor. For chaotic systems the boundaries of these basins of attraction may, again, have a fractal nature, showing a kind of irregularity that appears and reappears at every level of scale. This is, once again, because of the instability of such chaotic systems. Arbitrarily close to an initial condition that leads a system to approach one attractor there will be systems whose initial conditions lead them to a different attractor.

One of the major themes of chaos theory is the study of how dynamical systems that behave in a non-chaotic way for some value of a parameter in the equations governing the system (a coupling constant, for example, or some constant indicating a degree of friction) may become chaotic systems as the value of the parameter changes. Here, once again, the work goes back to Poincaré and his study of how periods of orbits might double suddenly upon a parameter change. A great deal has been discovered about the modes by which simple non-chaotic systems can become chaotic as a parameter reaches a certain critical value. Especially informative has been the realization that such routes to chaos may have a "universal" quality, extending over many systems that otherwise vary in their dynamical details.

As in the case of the qualitative dynamics of the singularities of dynamical systems and the study of their stability, chaos theory presents us with a discipline in which the explanatory goals and methods of dynamics take on a shape unlike that which occupied dynamics in the earlier centuries. Once again, we will no longer be asking the following typical question: For an individual system, structurally specified, given its initial position and velocity, what will be its position and velocity at some later time. And, once again, our attention will be focussed not on some individual trajectory but on the structure of a whole class of trajectories.

17.3 STATISTICAL MECHANICS

In qualitative dynamics, including the study of singularities, of stability and of chaotic dynamical systems, we find something of interest to the scientific methodologist. These theories are all, in a sense, straightforward applications of the fundamental dynamical theory. No challenge to the basic posits of that theory is being proposed, and no modifications of the theory are being attempted. But what we have seen is that, in trying to apply the theory to cases of particular interest, new and wholly unexpected structures of conceptualization and explanation can arise.

In statistical mechanics we see this methodological phenomenon once again. But in this new application of dynamics the motivation behind the theory, the kind of novel conceptualization proposed, and the explanatory structures unveiled are quite different from those encountered in qualitative dynamics.

This new application of dynamics arises out of thermodynamics, the theory that offers a lawlike characterization of the macroscopic behavior of systems that involves their thermal properties. The theory, which developed into a formal discipline in the nineteenth century, encompassed the realization that energetic exchanges of a system with the external world could take place either in the form of overt work or in the form of the flow of "heat," and that a conservation principle taking heat flows into account generalized the conservation of energy familiar from some cases (elastic collisions, for example) in pure dynamics. The theory also took account of the notable asymmetry in time of thermal processes, such as the flow of unevenly distributed energy into an equilibrium where all temperatures were equal. This is summarized in the form of the famous Second Law of thermodynamics.

That the flow of heat might represent the changes in the mechanical energy of the microscopic constituents of a system was an idea with its origins as early as the eighteenth century. But in the nineteenth century it finally took the form of detailed proposals as to how this "internal" energy of a system might exist. Most noteworthy was the kinetic theory of gases, in which a gas was posited to consist of microscopic, molecular components whose energy of motion was "the quantity of motion that we call heat" (Clausius).

Important work by Maxwell and Boltzmann established a proposed velocity distribution for such molecules when the gas was in a situation where its macroscopic condition remained stable over time, when it was in equilibrium. Both scientists also attacked the problem of how a gas not in equilibrium might approach that state. Partly on their own initiative, and partly in response to profound criticisms launched at the new theory, both Maxwell and Boltzmann realized that probabilistic concepts were essential to the theory.

A gas in a macroscopic condition has some dynamical condition, some set of positions and velocities, that is true of its molecular constituents. But to any one macroscopic state there may correspond innumerable microscopic conditions. The fundamental principle of what became full-fledged statistical mechanics was to propose a probability distribution that assigned to any gas in a given macroscopic state a probability that its microscopic state would be found to be in some specified range of phase space.

For the theory of systems in equilibrium a curious semi-autonomous theory developed, one that neglected all questions of how, dynamically, a system in non-equilibrium arrived at the equilibrium condition. One identified the macroscopic features of the gas, its pressure and temperature for example, with quantities calculated as the average value of some function of the possible dynamical microstates of the system of molecules that made up the gas. The justification for taking such average values as equilibrium values is a subtle one involving the very large numbers of molecules in the system and the special nature of the functions involved. Basically one tries to show that such average values will also be the overwhelmingly most probable values of the functions, and identifies these values as the ones an individual system would possess for an overwhelmingly great part of its lifetime as an isolated system. One then sought a probability distribution that would remain invariant in time as the states of each system of molecules in a collection of such systems all evolved according to the dynamical laws governing the systems of molecules. And one tried to show that such a probability distribution would be unique. One then argued that such a probability distribution and averages calculated with it would be appropriate to use to calculate equilibrium values because of the very definition of equilibrium as an unchanging state.

Finding a probability distribution over the microscopic states that was time-invariant proved easy. Showing that it was unique was more difficult. Here deep considerations of dynamics were needed, giving rise to a new discipline, ergodic theory, which is closely connected to the disciplines needed for the characterization of unstable systems discussed above. What one needed to show, it turned out, was that, for all but a set of initial conditions with probability zero, in a class of standard probability measures, the trajectory in phase space of the point representing any system would eventually go through any set of phase-space points of non-zero probability. This is called "ergodicity." A sufficient condition for this to be the case was found (metric indecomposability) and, after many decades, it was shown that an idealized system, closely related to the structure of a realistic gas, had the appropriate property. The idealized system was that of spherical particles in a box that interacted only by perfectly elastic collisions. The proof of ergodicity is closely related to the work that goes on in dealing with chaotic systems. Here again what is crucial is that trajectories from points initially close to one another diverge from one another with sufficient rapidity.

With ergodicity one can show that the standard probability distribution chosen to represent equilibrium is the only probability distribution, among those that agree with it as to which sets of points have zero probability, that is constant in time. And one can show such things as that, except for a set of systems of probability zero, a system will, over an infinite time, spend a portion of its time with its state in a given phase-space region equal to the size of that region in the standard probability measure.

Many profound questions remain open. Realistic systems probably are not ergodic. This is due to results similar to those of the KAM theorems. It is also far from clear why one is entitled to ignore sets of probability zero in the standard measure. Overall, autonomous equilibrium theory is curious in its neglect of the dynamical reasons for why systems approach equilibrium in the first place. Nonetheless, one can see here how, once again, a novel attempt at the application of dynamics has resulted in a conceptual structure quite unlike that found in ordinary theorizing in dynamics. New explanatory modes are employed to offer us explanations that are totally unexpected, and that are quite distinct from the usual modes of explanation found in traditional dynamics.

Both Maxwell and Boltzmann found equations that describe how the velocity distribution of the moving molecules would change from a non-equilibrium to an equilibrium condition. But they invoked probabilistic postulates that were not justified from the underlying dynamics and that might even have been inconsistent with that underlying dynamics. These postulates also contained implicitly the means by which the time asymmetry necessary to capture the Second Law of thermodynamics was imposed on an underlying time-symmetric dynamics. Going beyond their work required both conceptually rethinking the problem and developing new insights into dynamics.

One approach tries to show that there is implicit in the dynamics a process that will take a probability distribution that can be thought of as appropriately characterizing non-equilibrium and will change that distribution over time until it approaches a distribution that can, in some sense, be thought of as appropriate for characterizing equilibrium. This process is called "mixing." The arguments invoked in showing that some idealized system can undergo mixing once again resemble those used in characterizing chaotic systems, with trajectory instability providing the key to the deduction. As in the case of chaotic systems, it is a kind of macroscopic randomness superimposed on the underlying deterministic evolution at the microscopic level that is crucial. The initial probability distribution over microstates appropriate for non-equilibrium is said to approach the equilibrium distribution "in a coarse-grained sense."

Even when mixing is demonstrable, and there are the same problems here as in showing ergodicity in equilibrium statistical mechanics, one needs to supplement it with an appropriate constraint on the initial probability distributions in order to get the results one desires. Here both the issue of why and how probabilities become appropriately assigned to initial microscopic states of systems characterized in a macroscopic way and the issue of how the time asymmetry of the evolution is to be physically explained remain important open questions. The former issue raises questions of whether or not at some deeper level than the usual dynamics of the molecules some "physical randomness" exists in the world. The latter issue raises the question of whether the time asymmetry of the world is to be found in some deeper level of dynamics or whether, instead, it rests on some special initial state of the cosmos at the posited "Big Bang."

To further complicate the story, there is the fact that there exist alternative dynamical accounts of the approach to equilibrium from non-equilibrium that make no use of the "mixing" results at all. These alternative accounts find their resources, rather, in the vast number of molecules in the system and in such facts as the low density of the gas systems for which Maxwell and Boltzmann's equations apply.

In any case, non-equilibrium statistical mechanics once again reveals to us how an attempt at applying the fundamental dynamical theory in a new context calls up the need for novel conceptualization and novel explanatory patterns. Once again we have moved away from the traditional pattern of looking at the behavior of an individual solution to the underlying dynamical equations and away from asking, for a specific initial condition, how things will evolve. Once again it is a whole class of initial conditions that is invoked, and the evolution of that class in a qualitative and quantitative manner that is studied. Here the necessity of invoking probabilistic considerations often leads to the study of how a probability measure over initial conditions will evolve, driven by the dynamic evolution of each of the micro-conditions in the set of conditions to which the probability distribution is attributed. In still other treatments of non-equilibrium, considerations of a topological nature on the phase space, rather than measure-theoretic considerations, are used in conjunction with the dynamics to try to characterize the evolution toward equilibrium so familiar from the macroscopic world and the thermodynamic theory used to describe it.

Because the statistical-mechanical theory has as its goal an account of that macroscopic world and an explanation for the success of the theory used to characterize it, another important set of conceptual and explanatory issues comes into play here. The relationship between the theory at the macro-level and the dynamical theory used to account for the behavior of the microscopic constituents proves another rich field for methodological exploration. Here issues of the alleged "reduction" of one theory to another provide an additional level of complexity for us to deal with when the conceptual and explanatory aspects of dynamics as applied are explored in the case of statistical mechanics.

17.4 NEW APPLICATIONS AND EXPLANATORY PROJECTS

Each of the three fields of application of dynamics that we have just looked at – qualitative dynamics, chaos theory and statistical mechanics – required the invocation of new concepts not found in the previous dynamical theory. Each discipline also developed new objects of explanation, new kinds of behavior of systems that required explanatory accounts. And each new discipline developed new models of explanation to account for these newly discovered features of the world.

One can summarize these changes briefly and crudely, while acknowledging that such a summary can only begin to unpack the richness of these new conceptual schemes. The older applications of dynamics focussed primarily on methods for predicting the detailed future behavior of a system given its initial dynamical state. The new applications focus, rather, on surveying large classes of systems, each system of which has a different initial condition, as a whole. It is the characteristics of the whole class of systems that are of primary interest. Where do the singularities in the phase space exist and what are they like? What classes of trajectories remain stable under perturbations? How does chaotic motion develop out of non-chaotic motion when parameters are varied? What do the classes of chaotic motion look like in the set of all possible trajectories? How can probability distributions over whole classes of systems be assigned so as to characterize the equilibrium state found at the macro-level? How can such probability distributions be assigned to non-equilibrium states, and how do these develop over time, thus characterizing the approach to equilibrium?

One must be cautious here not to over-dramatize the differences between dynamics as traditionally applied and as applied in these new investigations. The new theories certainly aim for predictions of a quantitative nature, despite the plausibility of saying that there is an aspect to these new applications that might be called more "qualitative" than the earlier approach. And the new disciplines certainly offer many devices that are useful for predictive purposes, despite the fact that in many cases the detailed prediction of a future precise state for an individual system is no longer the primary goal.

One way in which these new applications provide an arena for methodological investigation is the way in which they demand, of their very nature, the application to dynamics of new, more abstract, levels of mathematical representation. Whereas traditional dynamics was almost entirely grounded in analysis, namely the applications of the methods of the calculus to the description of nature, the new theories demand the invocation of several abstract mathematical disciplines. Once again, one must not exaggerate here, for the discussions of symmetry of dynamical systems had already been leading to the application of abstract algebra to dynamical problems.

But still, the application of the methods of topology to qualitative dynamics, as in, for example, the Poincaré–Birkhoff topological proof of the existence of an infinite, dense set of periodic trajectories under very broad conditions in dynamics, constitutes one way in which new, abstract mathematics plays a novel role in dynamical theory. The invocation of measure-theoretic considerations, first in Poincaré's elegant proof of recurrence for broadly characterized dynamical systems and later, much more widely, in the application of probability theory to dynamics that is the core of statistical mechanics, is another example of how abstract mathematics, often developed in conjunction with the physics to which it was being applied, began to play a crucial role in the development of dynamical theory.

Statistical mechanics differs from the earlier applications of dynamics in another way that demands methodological inquiry. This new application of dynamics is designed to serve as the explanatory underpinning of an existing theory, thermodynamics, that has its own conceptual framework and its own explanatory structure in place. And these aspects of thermodynamics seem, initially, not to be of the sort that could fit into a dynamical framework at all. What do heat, temperature and entropy have to do, after all, with motion and force? There is much to be learned in studying just how the older thermodynamical theory and the newer statistical-mechanical theory are to be related to one another so that the latter can, in some sense, "account for" the success of the former.

The new applications of dynamics also require a careful examination of the way in which their application to the real world is mediated by the mechanisms of idealization. The older applications of dynamics required idealization as well, of course. Traditional dynamics dealt with point masses, perfectly elastic collisions, rigid bodies and viscosity-free fluids – none of which exist in nature. But the way in which idealization functions in the new applications is sometimes distinct enough to require special methodological examination on its own. In the case of chaos theory the way in which idealization is applied at the structural level, rather than to the individual motions themselves, is important. And in statistical mechanics the idealizations invoked – billiard-ball models of molecules, infinitely dilute gases with an infinity of molecular components, behavior of systems in the limit of infinite time – all play essential roles in the explanatory structure. Understanding how those explanations can play their legitimate role in accounting for the behavior of realistic systems under realistic conditions is a major task in the philosophical exploration of the foundations of statistical mechanics.

We will not be able to pursue in depth any of these fascinating problems here. Once again, there are three reasons for that. First, the central concern of this book is how reconstruals and reconstructions of the foundational principles of dynamics changed science and philosophy over the ages, not how these new applications of dynamics raised their own important methodological issues. Second, dealing with these issues in the depth they deserve would require several book-length works. Third, in several cases, chaos theory and statistical mechanics in particular, there exist substantial works in the literature devoted to just these issues.

Nevertheless, it would be appropriate here to make a few remarks about the place these applications of dynamics have in the general scheme of scientific theories, and about the problems they present to the methodologist concerned with the analysis of scientific theories.

It is a commonplace of methodological philosophy of science that special sciences exist along with the foundational theories of science. Even if we acknowledge some claim to universality of foundational physics, agreeing that in some sense everything that happens in the world is governed by its general principles, we may still claim many kinds of autonomy from that foundational theory for the scientific theories constructed to deal with their particular domains. There is no reason at all to deny that there is a place for autonomous principles of evolutionary biology, say, or for independent principles of cognitive psychology, just because we may agree that all of nature falls, in principle, into the domain covered by the foundational laws of physics.

Much attention has been given to issues of "irreducibility" or of "emergence." In these studies care has been taken to explain how there can be predictive and explanatory roles for principles of the special sciences, where these principles are in no way derived from the principles of foundational physics, and, indeed, where no way can be seen, even when the principles of the special sciences are known, for deriving those principles from those of foundational physics even as an afterthought. Much attention has also been given to the issue of the autonomy of the fundamental concepts of these special sciences.

Nothing in our acknowledgement that foundational physics has a claim to universality need lead us to think that there could be some useful sense in which the concepts of evolutionary biology or of cognitive science could be "defined" in terms of the concepts that play their role in foundational physics. And much attention has been devoted to the notion of "supervenience," the idea that the facts of the world dealt with by the special sciences are all, in some sense, determined or fixed by the facts of the world dealt with by foundational physical theory despite the autonomy from the concepts of foundational physics of the concepts used to characterize these facts in the special science and the autonomy of the laws of the special sciences from the laws of the foundational physical theory.

The disciplines used to characterize the modern applications of dynamics that we have been surveying – qualitative dynamics, chaos theory and statistical mechanics – have a curious place in the domain of all of our scientific theories. On the one hand, they too seem to have some degree of "autonomy" from the basic concepts and principles of dynamics as it had previously been understood. On the other hand, they are far more intimately related to the earlier dynamical theory than a special science is to fundamental physics.

In the case of the special sciences the situation is usually one where the concepts and explanatory schemes of the science have developed out of the necessity for dealing with a collection of observations and experimental facts in a manner that is more or less independent of any underlying, assumed foundational theory. When trying to develop the resources necessary for creating a theory of evolutionary biology or for constructing a theory of cognitive psychology, it is the observed phenomena of the development of the structure of organisms over time or the observed phenomena of human and animal behavior that constitute the guide to the desired theoretical structure, not, at least initially, any views about how fundamental physics describes the world. Of course, as these special theories develop it may become very necessary indeed to relate them to theories at a more foundational level. Evolutionary biology must come to grips with molecular biology (especially that of genetics), and cognitive psychology must be integrated with biological neurology. But the special sciences come first and their connections to deeper-level theories come later.

In the case of the qualitative dynamics of singularities and stability, the theory is being developed out of the underlying foundational dynamics. New concepts, global concepts framed in measure-theoretic and topological terms for example, are being introduced, but their introduction is clearly guided by and derivative of the older concepts of the foundational theory from the very beginning. Indeed, even to work within traditional dynamics requires such constant conceptual innovation. As dynamics became applicable to ever more complicated systems – rigid bodies, fluids, non-symmetric continuous media, for example – each new stage of application required the introduction of new conceptual elements. The notions of moment of inertia, viscosity and the stress tensor are examples of such newly introduced descriptive concepts, each bringing with it novel explanatory forms. The conceptual innovation needed to do modern qualitative dynamics is, to be sure, one which adds to the existing theory in a more dramatic fashion than did the earlier innovations. But it is all a matter of degree.

On moving to chaos theory, one sees some additional elements in the conceptual innovation, the use of the measure-theoretic devices in order to construct more extensive probabilistic arguments, for example. And, of course, in the case of statistical mechanics the changes are more striking still. Here we have a situation that looks a bit more like the standard cases of the special sciences. Thermodynamics was such a special science, with its fundamental concepts of temperature and entropy and its fundamental laws, especially the Second Law, introduced in a manner that was quite independent of any views about the underlying mechanical constitution of macroscopic media or the dynamics that governed that constitution.

But statistical mechanics itself is, once again, a discipline that takes its very being from the concepts and structures of the underlying dynamics. The basic concepts of classical dynamics and the basic dynamical laws are always to the fore in each conceptual construction and each explanatory pattern of statistical mechanics. Here again, though, much is added to the earlier versions of the dynamical theory. In particular the use of measure-theoretic concepts to characterize probabilities, and, in some cases, topological concepts to characterize other versions of the sparseness and denseness of systems, goes far beyond anything present in the earlier dynamical theory. This gives rise to many of the most fascinating foundational questions about statistical mechanics. Can these probabilistic notions be themselves "grounded" in the underlying dynamics? Or do they require new posits that are independent of the earlier foundational laws? In particular, how is the temporal asymmetry of the use of probability in the theory to be reconciled with the underlying temporal symmetry of the dynamics? Here we have a situation where a new theory is far more deeply integrated in its origins and in its structures with the foundational theory than is any one of the truly special sciences. But, at the same time, it is a theory that goes beyond the foundational theory itself in ways that suggest some degree of autonomy for its concepts. Indeed, a careful exploration of the theory has suggested to some that many of the basic features of the theory are grounded in thermodynamic notions rather than in pure dynamics, a claim that makes the issue of the "reducibility" of thermodynamics to dynamics one that is deeply problematic.

Suffice it to say that the existence of these rich and extensive structures, the theories necessary to apply the underlying foundational dynamics in new ways never guessed at by those who developed the earlier theory, provides a new field of exploration for the methodologists. These new conceptual and explanatory structures are not derivable from the earlier theory in ways that are transparent. Even when the new results do follow from the earlier principles, it is only by the introduction of new conceptual schemes and new mathematics that this can be seen. And in some cases, say the postulation of initial probability distributions in non-equilibrium statistical mechanics, it is likely that wholly new independent posits are called for. Yet these new structures for understanding the world are not like the autonomous special sciences either. Both in genesis and in structure they are far more intimately connected to the earlier foundational theory than are the true special sciences. These "application theories" of dynamics are a component of theoretical science in their own right and require their own methodological understanding.

SUGGESTED READING

An excellent non-technical introduction to celestial mechanics and the stability problem is given by Diacu and Holmes (1996), which covers Poincaré's "mistake," singularities in dynamical systems and the modern stability results of the KAM theorem. A nice historical discussion of Poincaré's discovery is given by Barrow-Green (1997). Sternberg (1969) explains the modern approaches to celestial mechanics. For an extensive treatment of the older approaches to celestial mechanics, see Winter (1947). Devaney (1986) gives a very clear introduction to chaos theory. The book by Moser

(1973) is a classic introduction to the solution to the stability problem in celestial mechanics. Smith (1998) gives a very clear and sound discussion of chaos theory from a philosophical perspective. The role of dynamical instability in statistical mechanics is surveyed in Chapters 5 and 7 of Sklar (1993).

CHAPTER 18

Spacetime formulations of Newtonian dynamics

The theory this book is exploring, Newtonian dynamics, is, of course, a false theory of the world. Several scientific revolutions have shown us that it must be replaced as our fundamental theory of motion and its causes.

The special theory of relativity rejects even the basic kinematics of the Newtonian account. Even at the level of the very description of motion, at the level of our construal of the spatial and temporal structure of the world, the Newtonian account is rejected. At the level of dynamics, the level at which the causes of changes of motion are explored, the theory, once again, replaces the Newtonian account with a novel theoretical structure. The general theory of relativity proposes even further, highly dramatic, changes in our very notions of space and time. Its most direct contact with the Newtonian picture of the world is to replace Newton's famous account of the origin, nature and effects of gravitational force with a new account of the gravitational interaction of the material of the world.

The quantum-mechanical revolution forces even more dramatic revisions of the Newtonian world picture. Here the most basic concept of Newtonian theory, the very idea that systems have determinate positions and states of motion, has been challenged by some interpretations of the new theory. Once again, both the Newtonian kinematical picture and the Newtonian dynamical theory are replaced by wholly novel structures for characterizing our world.

One set of questions of profound philosophical interest is how the novel theories are developed by a process that utilizes the fundamental structures developed for the Newtonian theory, but which subjects those structures to vital processes of transformation in order to find the new structures needed for the new theories developed in the light of experimental results incompatible with Newtonian physics.

Another set of questions also of profound philosophical interest deals with the relationship between the Newtonian theory and each of its successor theories. Here issues of whether or not Newtonian theory can be viewed as some kind of approximation to the novel theories, and whether or not one can speak of the Newtonian theory "reducing" to the novel theories by means of some "limiting" process or other, are typical matters of concern.

But these questions are, fundamentally, really questions about the novel theories and their deep structures. They are, in any case, beyond the scope of this book. The issues here have received much attention in the physics and philosophy literature, but much more work remains to be done in order for them to receive their due.

In this chapter we will explore an issue of a different kind, one that is essential for our main purpose of looking, philosophically, into the structure of the Newtonian theory itself. The two relativistic theories each require a novel spatio-temporal setting for their dynamical theories. These are the Minkowski spacetime of the special theory and the curved, pseudo-Riemannian, spacetime of the general theory. Each of these spacetime accounts starts off with the novel idea that the "points" of the spacetime are to be taken not as places or instants, but as "event locations."

The work of a series of brilliant mathematicians of the decade subsequent to the introduction of general relativity revealed that one could deeply illuminate the now rejected theories of Newtonian dynamics and Newtonian gravity by reconstructing the earlier theories using devices that had been discovered to be essential to formulate the new, relativistic, theories. Here we have a process of some methodological interest: Devices generated only in the context of formulating new theories turn out to be essential for a retrospective understanding of the foundations of the theory they have replaced. The procedure has subsequently been of value in other interpretive programs as well. For example, the novel devices needed to understand the role of the electromagnetic potential in the quantum context have proven of deep value in understanding what that potential was even in the earlier, pre-quantum, context of Maxwell's original theory.

It is to this process of retroactive reinterpretation of Newtonian kinematics and dynamics in the light of relativistic spacetime theories that we now turn. After first briefly outlining some of the structure of the spacetimes of special and general relativity, and just a very little of the motivation behind the adoption of these structures in the new theories, we will explore how a flat spacetime does better justice to Newtonian kinematics and dynamics than did Newton's "absolute space" and "absolute time," and how a curved spacetime provides a better framework for Newton's gravitational theory than did Newton's own. Then we will reflect a little philosophically on what these reconstructive moves do, and do not do, to resolve the long-standing questions about the place of space and time in the Newtonian world-picture.

18.1 RELATIVISTIC SPACETIMES

The spacetime of special relativity, Minkowski spacetime, was developed to provide a natural framework for Einstein's new proposed laws of nature.

Crudely, the aim was to find a way of describing the world that admitted the laws of electromagnetism, laws which stipulated a determinate value for the speed of light, with this speed uniform in all directions, but which held to the old Galilean principle that physical behavior looked the same to all observers in one of the preferred, uniformly moving, inertial reference frames.

It was Einstein's genius to see clearly that this required a new notion of simultaneity for events at a distance from one another, a notion that relativized which events were or were not simultaneous to the choice of inertial reference frame. Allowing such relativization, and sustaining the principle that light had the same fixed speed, uniform in all directions for all inertial observers, and then embedding the results in the simplest possible spacetime compatible with such posits resulted in the spacetime first described by Minkowski. The introduction of this new spacetime required also a rejection of Newtonian dynamics and the construction of a new relativistic kinematics and dynamics.

Here we will not follow the familiar construction of that spacetime, since it is Newtonian, rather than relativistic, dynamics that is our concern. But one crucial move in Minkowski's program will be essential for our purposes. That is the choice of the basic "points" of the spacetime as locations of possible infinitesimal events, events such as the collision of pointlike particles or the intersection of one-dimensional light rays. These event locations, rather than places or instants, are the irreducible elements of which the spacetime is constructed. Following a procedure that had become familiar from pure mathematics, one could build up the full structure of the spacetime by imposing on the set of basic elements one after another of the needed structural features. In the relativistic case one ended up with a spacetime with a definite metric structure to it, but one which, unlike that of ordinary distance, was such that the square of the "interval" between points need not be non-negative definite. In addition the interval between events could be zero even for distinct events. The spacetime remained, however, flat, in a rigorous mathematical sense.

Curvature was introduced into the spacetime of a physical theory by Einstein in his search for a new theory of gravitation. It was plain that the existing Newtonian theory was incompatible with the constraints imposed by special relativity. Many approaches could be, and were, taken in the search for a relativistically acceptable replacement theory of gravity. Characteristically, however, Einstein's proposal was the most radical, approaching the new theory from deep first principles.

Curved spaces differ from flat spaces in several ways. First there are the metric aspects. Distance relationships among points on, say, the curved surface of the Earth are quite unlike the relationships among distances sustained on a two-dimensional flat, plane surface. But curved surfaces also fail to have straight lines as both lines of least curvature and lines of shortest distance between points. Instead, generalizations of the straight line, straightest possible curves and shortest-distance curves, geodesics, take the place of the straight lines of the flat spaces.

Several heuristic arguments led Einstein to consider curved spacetime as the appropriate means for formulating a relativistic theory of gravity. Some of these dealt with the metric aspects of spacetime. All of these relied upon the observation that the simple mechanical situation of an object falling in a uniform gravitational field resulted in motions that could be duplicated by eliminating the gravitational field and instead observing the inertial object from an accelerated reference frame. Einstein guessed that such an "equivalence" could be held to in general as a guiding principle for a theory of gravity. Relativistic effects of accelerating frames on time and space measurements then led to the idea that gravity should have such metric effects, suggesting that gravity might be thought of as being a change of spacetime from a flat to a curved manifold.

More relevant for our purposes is a simple reflection of the dynamics of matter in a gravitational field. It was Galileo who first pointed out that the rate of fall of objects in a gravitational field is independent of their size and of their constitution. Newton realized the deep importance of this observation and, indeed, tested it with his important pendulum experiments. This fundamental fact about gravity becomes embedded in the Newtonian theory by having the gravitational charge that appears in the formula for gravitational force, and that mediates both the force one particle exerts on another and the force that a particle feels from the other, be nothing but that same intrinsic inertial mass that measures an object's resistance to acceleration when suffering a force.

The result is that in the gravitational field of an object each and every object with the same initial position and velocity will follow the same trajectory due to the gravitational field. It is this fact that allows, for example, the astronomer to calculate the mass of Jupiter from the motion of its moons, since these motions depend not at all on intrinsic facts about the moons but only upon facts about Jupiter and observable initial conditions for the satellites.

It was Einstein's genius to see that this fact suggested a possible geometric reading for the nature of gravitation. In a curved space one and only one geodesic path goes off from a point in a given direction. In spacetime a point is an initial position at an initial time. An "initial direction of motion" encompasses both the spatial direction of the initial motion and the initial speed of that motion. Then an initial position and direction will determine a unique (timelike) geodesic in the spacetime. But since all particles started at certain a place at certain a time with the same initial velocity have the same following trajectory when subject to gravity, it becomes heuristically natural to think that one might be able to identify the path of an object in a gravitational field with the geodesics of the spacetime, a spacetime now thought of as "curved," since "free" paths in a gravitational field are assuredly not straight spacetime paths.

Heuristic arguments are not, of course, constructions of a specific theory and not demonstrations of empirical plausibility for such a theory. Einstein's great accomplishment was the construction of a theory, general relativity, that did justice to the heuristics, constituted a relativistically acceptable theory of gravity, and has subsequently received serious empirical confirmation.

The theory has a spacetime that is curved. It has a metric structure that generalizes that of Minkowski spacetime by having a pseudo-Riemannian metric. The flat spacetime of Minkowski is retained only as a local good approximation to spacetime at any point, that is, as the geometry of the tangent spaces of the general relativistic spacetime. The key is that observers freely falling in a gravitational field will, at their location, find local physics described by special relativity. But putting the results of all observers together can be done only in a curved spacetime. In this curved spacetime the freely falling material test objects all follow the curved, timelike geodesics of the spacetime. Beyond this, general relativity's other key posit is a new law connecting the distribution of ordinary mass-energy in the spacetime to the curvature structure of the spacetime. This law replaces the old Newtonian formula for finding the gravitational field generated by a particular distribution of attracting matter.

18.2 GALILEAN SPACETIME

Some time after all of this became familiar, it was realized that the basic move in the Minkowski construction, namely picking "events" (or, better, the locations of possible, infinitesimal events) as the basic entities and building the spacetime by imposing structures on the set of events, could be used to retroactively construct a spacetime for the now-rejected Newtonian kinematics and dynamics. This new construct, it could be argued, was far better suited for Newton's experimental data and for Newton's theoretical account of that data than was Newton's own "absolute space." The new spacetime has been called "Galilean spacetime" and also "neo-Newtonian spacetime." What is it like?

Time is taken to be just as Newton thought, absolute. There is a definite time interval between any two events, unique up to the possibility of arbitrarily picking a zero point for time and an arbitrary scale factor. But there is no "relativity" of time interval to reference frame as there is in special relativity. Time intervals are invariant. And, in opposition to some versions of relationism, there is a preferred metric for time and not any old functional variation on absolute time is allowed.

Consider a set of events all of which occur at the same time, that is, all of whose pairs of events have a zero time interval between the members of the pair. It is posited that such a class of event locations forms an ordinary, three-dimensional Euclidean space, with all of the standard metric features of such a space true of the spatial distance relationships among the simultaneous events.

Newton believed in "absolute rest." An object at absolute rest occupied "the same" spatial position through time. But this makes sense only if we can say of two non-simultaneous events whether or not they are at the "same spatial place." Galilean spacetime does not have any such structure. More generally, there is nothing in the spacetime that corresponds to the notion of the amount of spatial separation between two non-simultaneous events.

But accelerations are changes in velocity over time, and velocities are changes in position over time. So how can the Newtonian notion of absolute acceleration exist in a world in which there is no such thing as how far an object is at one time from its place at another time? Of course *relative* velocities and *relative* accelerations still exist in Galilean spacetime, but these are not sufficient, as Newton so clearly argued, to allow one to define the needed absolute notion of acceleration. How can a place for absolute acceleration be found without introducing absolute place or absolute velocity?

The trick is to add one piece of needed structure to those already imposed on the set of event locations. One way to do this would be to posit a three-place relation on non-simultaneous events, the relation of the events being "collinear." So any triple of non-simultaneous events is either, absolutely, collinear or it is not. From this it is easy to think of an infinite set of non-simultaneous events as either constituting a "timelike straight line," or not so lying along a timelike straight line, in the spacetime.

The idea is to put into the spacetime structures that are empirically discernible in Newtonian dynamics and to leave out of the spacetime any structures that have no empirical correlate. Newton himself was well aware of the fact that his theory entailed, making certain assumptions about the nature of forces, that no dynamical observation could distinguish one inertial frame as the rest frame. The best he could come up with was the dubious suggestion that the center of mass of the solar system could be considered at rest, a suggestion that becomes absurd in the light of modern cosmology. But, he was at pains to emphasize, absolute accelerations were empirically determinable, this being the brunt of the famous argument in the "Scholium" of the *Principia*.

In Galilean spacetime there is no rest frame to be discerned. Inertial spacetime paths exist, but these have their empirical correlate in the motions of bodies not suffering forces. Identifying which particles are not suffering external forces is itself problematic from a conventionalist standpoint, but taking isolated objects moving distant from all other objects as force-free counts as legitimate within the Newtonian theoretical framework. And once we know how to determine that an observer is moving inertially, that is, without any "absolute acceleration," we can then determine the absolute acceleration of any object. This can be determined experimentally as just the relative acceleration of that object relative to any chosen inertial reference frame.

So in Galilean spacetime there exist only those kinematic and dynamical structures which play an empirical role in the Newtonian theory. We will reflect shortly on the meaning of all of this for an interpretation of the theory, but there is no question that, as a legitimate scientific step, the movement from Newton's spacetime with its otiose absolute rest to Galilean spacetime as a more appropriate spacetime background for Newtonian theory is the right methodological move.

18.3 CURVED GALILEAN SPACETIME AND NEWTONIAN GRAVITY

There is nothing in the Newtonian theory of gravity that would suggest the kind of metrically curved spacetime that functions in general relativity to be relevant to the Newtonian account. Indeed, since we are trying to reconstruct a Newtonian world, we should try to adhere to most of the Newtonian temporal and spatial notions. Absolute time must remain absolute in the Newtonian sense; and spaces, sets of simultaneous events, must remain the standard, flat, three-dimensional space of Newton.

But the heuristic dynamical fact that was part of the inspiration for general relativity, the fact that gravity acts on material test objects independently of their size and their constitution, has nothing to do with the relativistic aspects of spacetime. It is a fact central to Newton's own theory of gravitational force, as we noted above. Can we find some spacetime way of embedding this fact in the Newtonian theory in such a way as to give the new account some of the same kinds of advantages over Newton's account of gravity that Galilean spacetime has over Newton's absolute space and time?

Here is one way to think about this: Newton's absolute space and time suffers from the problem that it posits a unique inertial frame as the frame of absolute rest. But the theory itself tells us that no empirical test can reveal to us which inertial frame is the rest frame. Galilean spacetime, by eliminating the rest frame altogether, removes this difficulty from the theory. At the same time Galilean spacetime retains empirically detectable absolute acceleration as part of its structure. Is there some similar problem with Newton's gravitational theory? And is there a similar spacetime way of obviating the difficulty? The answer to both questions is "Yes."

Newton explicitly realized that if a system was falling freely in a uniform gravitational field its internal dynamics would be just as if the system were inertial, or, indeed, at rest in absolute space. At least this is true if, as is usually assumed, the internal forces the parts of the system exert on one another are independent of the acceleration of the system. Newton needed this result to assure himself that he could treat the system of, say, Jupiter and its moons as if that system were at rest in an inertial frame. The system is not, for Jupiter and its moons together are falling in the Sun's gravitational field. But the distances of the moons from Jupiter are relatively small enough, compared with the planet's distance from the Sun, that one could think of the whole Jovian system as falling in a uniform gravitational field of constant force and direction at all points. Then, as Newton persuasively argued, no harm would be done by treating the system as being at rest in an inertial frame or in absolute space itself.

As Maxwell and others realized, this feature of Newtonian dynamics leads to a puzzle. Suppose that the entire Universe were permeated by a uniform gravitational field. All our empirical observations would be the same no matter what the uniform force value of this field and no matter in what direction it pointed. A similar paradox plagues Newtonian cosmologies with a Universe uniformly filled with matter. All observers' instruments record them in inertial motion in such a world. Each observer can explain this by positing itself as being at the center of the Universe with all the other observers accelerating toward it. The other observers' zero results on their accelerometers would be explained by the fact that all parts of their accelerometers were equally accelerated by gravity, the whole system again being in free-fall.

Once again a spacetime account of gravity can eliminate these "paradoxes." The spacetime is the curved version of Galilean spacetime. Once again the set of basic elements is constituted by the point event-locations. There is an absolute time separation between any two events. Sets of all simultaneous events form spaces that are three-dimensional Euclidean, flat space. The set of straight-line timelike paths that characterized flat Galilean spacetime is dropped. Instead there are preferred curved timelike paths consisting of events none of which are simultaneous with one another. These are to be associated empirically with the free-fall paths of material test particles.

These paths are the timelike "geodesics" of the spacetime. The mathematics of the spacetime is, alas, a bit trickier than that of general relativity. The relativistic theories have the speed of light as a scale to compare spatial and temporal separations. This is what allows them to have not just the relative spatial and temporal separations of events, but an invariant spacetime interval separation between them. Without such an invariant speed, and without the determinate spatial separation between non-simultaneous events that Newton's absolute rest frame would provide, neither flat nor curved Galilean spacetimes have metric structures for the spacetime as a whole. There are invariant temporal separations between events, and there are invariant spatial separations between simultaneous events, but there is no determinate spacetime interval between non-simultaneous events.

Nonetheless, there are the determinate curved paths followed by "free" particles, that is to say, by particles influenced only by gravitational forces. These are, once again, determined by initial spacetime points and initial spacetime directions, and are independent of the size or constitution of the particle chosen. So we can place them into the spacetime structure as is done in curved Galilean spacetime.

The result is a spacetime background for Newtonian dynamics with gravity that eliminates the distinction between empirically identical worlds, such as material universes falling in distinct uniform gravitational fields, but that preserves the empirically determinate structures of the Newtonian account. In this way curved Galilean spacetime does for Newtonian gravitational theory what Galilean spacetime did for Newtonian dynamics when it abolished the existence of an absolute rest frame but preserved absolute acceleration.

18.4 PHILOSOPHICAL REFLECTIONS

The reappraisals of the appropriate spacetime setting for Newtonian dynamics and for Newtonian gravitational theory that are provided by Galilean spacetime and curved Galilean spacetime are fascinating for several reasons. First, these reconstructions of Newtonian theory are of great interest methodologically when the issue of the very nature of an "interpretation" of a theory is in question, and when one puzzles over what resources are to be taken into account when one is engaged in the process of interpreting a theory. Second, the particular structure of the reinterpretations of Newtonian theory by these new spacetime frameworks is of interest for the general insights they provide into the "metaphysics" of interpretations and into the grounds that play a role in deciding which metaphysical understanding of a particular theory is "best."

Let us first look at some of the methodological issues, and then at some of the metaphysical aspects of interpretation.

The idea that theories are in need of interpretations is one that has never been thoroughly explored in detail, despite all the deep physics and philosophy that has been devoted to the specific possibilities for interpretations of individual fundamental physical theories. In order for there to be conceptual room for interpretation, one needs to assume that the usual presentation of a theory, namely the exposition of its mathematical apparatus and the usual material that appears in scientific presentations that connect that apparatus to possible experimental procedures and their outcomes, leaves open unanswered questions about the sort of world the theory ought to be taken as describing. The task of the interpreter, then, is to bring to light these hidden questions about the nature of the world for which the theory serves as partial, adequate description. And the interpreter should then go on to offer one or more possible answers to these questions.

The idea that theories can have multiple interpretations offered for them also presupposes that we can make sense of each of these interpretations being "understandings" of one and the same theory. Presumably offering one interpretation or another of a theory is not supposed to be the same thing as offering one or another genuinely alternative theoretical account of the world. This idea of many interpretations for one and the same theory is not an easy one to make clear. One initial obvious thought is to suggest that all the various interpretations of a theory must lead to the same empirical consequences if they are to be interpretations of one and the same theory. But, first, there are problems with the very notion of "all the same empirical consequences" for two distinct interpretations. Second, there is the radical positivist to contend with, who would argue that any such empirical equivalence reduces all the alleged interpretations to variant expressions of one and the same account of the world. Finally, there are cases, the GRW interpretation of quantum mechanics for example, where we are inclined to speak still of an "interpretation" despite the fact that the suggested account entails novel empirical predictions that are not present in the usual formulation of the theory being interpreted. We can hardly delve deeply into these issues here. We can, though, look briefly at how these new spacetime interpretations of Newtonian theory fit into the general pattern of interpretations of theories.

First we note that these new interpretations of the Newtonian theory are meant to leave all of the Newtonian observational predictions alone. They are versions of the theory that are intended to remain true in every way to its empirical predictions, and even, in some sense, to make no substantial modification to the "scientific" part of the theory. Their revision concerns only what is taken by the revisers themselves to be only the "interpretive" part of the Newtonian theories. When we explore shortly various Machian interpretations of Newtonian dynamics, we will see examples of the other kind of interpretation, those that suggest serious internal revision to the basics of the theory and even, perhaps, possible modifications of its experimental predictions.

The revisionary interpretations of Newtonian theory we have just outlined are interesting exemplars of a general theme. This is the methodological understanding that the job of understanding a theory from an interpretive perspective may require a grasp of the broader theoretical context into which the particular theory in question is embedded. Here the understanding of Newtonian dynamics and of Newtonian gravity is enriched by the use of conceptual apparatus designed not for the Newtonian theory, but for the successor theories of special and general relativity. It was only in the new scientific context that the spacetime concepts were developed that could be so illuminatingly retrospectively applied to the earlier theory.

One way in which looking at a broader theoretical context can aid us in developing an interpretation for a theory is to compare the theory in question with other physical theories that bear to it some appropriate analogy in terms of structure. We might, say, reflect on the possibility of a field-theoretical rendition of gravity in classical physics when the field-theoretical interpretations are developed for electromagnetic force. But, in the case we are looking at, the situation is rather different, for it is interpretation designed for successor theories, theories that replace the theories in question, that provides the insights into the nature of their now-rejected antecedent theories. This case is not unique in science. Consider, for example, the case explored by Gordon Belot of the way in which the gauge field interpretation of electromagnetism, which was developed in the light of the novel consequences of electromagnetism that show up only in the quantum-mechanical context, can be used to invigorate our understanding of the interpretation of classical electromagnetic theory. It is possible, then, for a theory to be "fully understood" only in the retrospective sense that its best interpretations come to mind only in the light of a theory that has replaced the one in question in our scientific corpus.

This particular set of reinterpretations of existing theories is interesting for another methodological reason. In discussions of the confirmation of theories the question of the role to be played by "hypotheses not yet imagined" often arises. How can we construct any kind of adequate theory of how data ought to be used to select which of a given set of hypotheses ought to be selected as "most believable," if we must keep in mind the fact that there are innumerable hypotheses that might account for the data that aren't even in the set of hypotheses being considered? Might not one of these "unconceived" (or "unborn"?) hypotheses be a better account of the data than any of those yet imagined?

But, it is sometimes asked, are there any real episodes in science where it turned out that, long after the theoretical decision making relevant to a particular theory had taken place, new imagination had provided some new hypothesis, arguably better to account for the data than any of those considered at the time the theory was originally adopted against its then known competitors? Well, here are two cases where exactly that is the case. Every methodological reason seems to argue that, had Newton been aware of Galilean and curved Galilean spacetimes, he would have thought them better interpretive accounts for Newtonian dynamics and Newtonian gravity than his own absolute space and time and his own gravitational force in flat spacetime.

What do these two reinterpretations of Newtonian theory tell us about some of the metaphysical and epistemological issues involved in the general notion of interpreting a theory?

These two reinterpretations are almost ideal cases of a general kind of interpretive program that appears again and again in trying to deal with difficulties in the understanding of fundamental physical theories. We are presented with a theory that seems to require the positing of unobservable theoretical structure and that seems to tell us that, by the theory's own account of the world, we must remain forever in ignorance about facts about this structure. Newton's interpretation of dynamics tells us that there is an absolute rest frame for motion but that there is no empirical means of determining which of the inertial frames is that rest frame. Newton's interpretation of gravity tells us that there is a fact about the magnitude and direction of the constant universal gravitational field of the Universe, but that there is no empirical way of determining what the magnitude or direction of this force is.

Both novel interpretations work in similar ways. First they assume that it is possible to declare certain aspects of the world spoken of by the theory as being forever outside the realm of direct empirical determination. They assume that there is no such thing as a "direct observation" that could tell us without inference what the rest frame is or what the universal uniform gravitational force is. Reflection shows us that the epistemic presupposition underlying the interpretations is that facts about "spacetime itself," and even some facts about the relations material things bear to one another, are outside the realm of "direct" determination.

Both interpretations, on the other hand, allow certain facts about the spatio-temporal relations among the particles of material to be taken for granted as open to epistemic access. The critique of Newtonian dynamics, for example, takes it that we can determine which particles are "moving freely" or are "unforced" in their motion, so that we can determine the inertial reference frames. It presupposes also that we have access to the appropriate standard for absolute time. As we shall see, the Machian interpretational program is critical of both of these assumptions. The critique of Newtonian gravitational theory presupposes that we can at least determine of the motions of material things which of them are free of all but gravitational force.

With these epistemic assumptions in place, both interpretations look for accounts of the world that will agree with the standard Newtonian account in all their empirical predictions. What is legitimately counted as being a prediction of the theory to be thought of as "empirical" is now determined by the epistemic presuppositions about what is and what is not to be supposed within the realm of the "directly observable." And it is then possible to demonstrate the "empirical equivalence" of the original Newtonian interpretations with their new spacetime substitutes.

The new interpretations are then suggested as being improvements over the older versions of the theory in that they have, to at least some degree, "thinned out" the postulated ontology of unobservables posited by the earlier interpretations. Although Galilean spacetime retains Newton's absolute time and absolute distinction between inertial and non-inertial motion, it dispenses with the Newtonian idea of an absolute rest frame, that is, with the idea that one could characterize non-simultaneous events as genuinely being, or not being, at the "same absolute place." Along with the rest frame goes any absolute notion of velocity with respect to "space itself." Curved Galilean spacetime goes one step further, dispensing with Galilean spacetime's structure of global inertial reference frames and replacing these with the structure of the timelike geodesics followed by particles moving "freely," that is, moving uninfluenced by anything except their own inertia and the gravitational field. Along with the notion of global inertial reference frames, the idea of the value and direction of any universal gravitational field of force is discarded.

In both cases the overall program is the same: Reconstruct your understanding of the theory in such a way as to try to eliminate from the posits of the theory the postulation of entities and properties characterized by features that, on the basis of the theory's own presuppositions and postulates, are not available for empirical determination. This overall motivation has been, of course, at the core of relationism ever since spatio-temporal relationism was first espoused by Descartes and Leibniz. But it is essential to realize that these two reinterpretations of Newtonian theory are certainly not yet relationistic accounts of the world. For they both still posit "spacetime itself," which is rich with its own structure. And it is the nature of this spacetime that reveals itself in its causal consequences for the observable facts of the world. Spacetime itself exists for Galilean spacetime and curved Galilean spacetime as it did for Newtonian space and time. There is a spatio-temporal structure of the world that exists over and above the spatio-temporal relations material things bear to one another. And the existence of this structure is a crucial element in the explanations we are to offer for the dynamic structure of motion and its causes.

In the following two chapters we will explore other aspects of the interpretation of Newtonian dynamics that are relevant in other ways to the old debate between substantivalists and relationists.

SUGGESTED READING

A clear exposition for philosophers of the use of spacetimes to reformulate Newtonian theory is given by Stein (1967). Another treatment is given by Earman (1970). Further expository treatments are to be found in Chapter III of Sklar (1974) and Chapter 2 of Earman (1989).

CHAPTER 19

Formalization: mass and force

The realization that a branch of knowledge could be presented in a form in which the entire contents of the field of investigation could be expressed by positing a small number of basic truths and by claiming that all the other truths of the discipline followed from these basic posits by pure deductive reasoning alone predated any serious development of dynamics or of other branches of physical science. The axiomatization of geometry has its origin at such an early date, in fact, that we have no good record of when or how the very idea of presenting geometry as a deductive formal discipline arose.

This early discovery of a branch of mathematics as a formal science had many consequences for the history of science and the history of philosophy of science. The entire history of the rationalist approach to knowledge in philosophy is founded on the early discovery that geometry could be structured as a set of consequences logically deducible from apparently "self-evident" first principles. Closer to our concerns, it is clear that Newton's *Principia* is itself structured to resemble as closely as possible the standard presentation of geometry. But there is, of course, no pretence on Newton's part that his first principles could themselves be established without reference to empirical experiment. (It was left to Kant to fall into the trap of trying to establish Newtonian dynamics as a fully a-priori science!)

At the end of the nineteenth century the formalization of mathematics by axiomatization became once more an active program. Many of the abstract systems of algebra (group theory, etc.) could easily be presented most perspicuously in axiomatic form. Hilbert revisited the alleged axiomatization of geometry familiar since Euclid and, for the first time, presented the theory as a rigorously axiomatic system whose first principles were complete and were independent of one another. In the foundations of mathematics the paradoxes uncovered in naïve set theory led to proposals for axiomatizing the notion of a "set" and to the long-term exploration of the possibilities for adding additional axioms to the existing axiom sets. The fear that even these axiomatic systems might contain implicit self-contradictions led to Hilbert's proposals for consistency proofs by way of meta-mathematical investigation and, ultimately, to the great discoveries about the very limits of formalizability of mathematical disciplines springing from the work of Gödel.

It was Hilbert who posed, as one of his famous challenges to future mathematics, the formal axiomatization of the foundational theories of physics. At the time of his proposal it would have been Newtonian dynamics that was the central challenge for such axiomatization. Here we shall look at two programs designed to present Newtonian dynamics in a formal mode. The degree of formalization on which they insist is quite different. And the purposes for which formalization is being sought are also distinctively different. We shall first look at a program of "informal formalization" whose aim is to provide a comprehensive mathematical framework in which to embed Newtonian dynamical theory in its most difficult realm of application, namely continuum mechanics. Then we will look at a different program, one that demands far more rigid criteria of formal rigor. This program has been carried out only for the simplest application there is of Newtonian dynamics, the dynamics of a finite set of point masses. Here working through the analytical subtleties of applied mathematics is not the goal. The aim is, rather, to find a means of clarifying the long-standing conceptual issues about the role of such concepts as those of "mass" and "force" in the theory. As we shall see, significant progress on those questions can be aided by the formalizing program. But the deepest philosophical puzzles about mass and force remain unresolved by the formalization program.

19.1 "INFORMAL" FORMALIZATION

The basic principles of dynamics are, by themselves, empty in their predictive content. It is only when dynamics is applied to specific systems, systems whose components exert specific forces and torques upon one another or whose energetic functions take on particular forms, that the predictive capacity of the theory is realized. There is a very wide range of kinds of systems to which we desire to apply the theory: Finite sets of "point" particles, rigid bodies, cables and chains, simple fluids, viscid and compressible fluids, complex continuous bodies with complicated non-linear responses to applied forces, and fields with non-denumerable degrees of freedom are all fair game for dynamical description and prediction.

Each kind of application of dynamics presents its own special difficulties, and each generates its own special methods for overcoming those obstacles. For finite sets of point particles, for example, framing the dynamical equation is easy, but solving it, even for almost the very simplest cases, is very hard indeed. Hence the elaborate resources of celestial mechanics and the deep methods for solving many-body problems. Hence, also, the development of a profound theory of successive approximative solutions. Some applications cry out for the "energeticist" methods developed in the eighteenth and nineteenth centuries. Motions of systems under "nice" constraints, say holonomic constraints where the forces of constraint do no work, beg for the methods of Lagrange and Hamilton. The dynamics of fields, again, is almost entirely framed in terms of Lagrangian methods. Here the dynamics derived from d'Alembert's principle or from extremal methods finds its greatest use. On the other hand, if one is using the dynamics of molecules to underpin a probabilistic account of thermodynamics as in statistical mechanics, it is the phase-space representation of the theory and the resources provided by the integral invariants of Hamiltonian dynamics that are central to the application.

When dynamics is applied to complex problems in the behavior of continuous media many special difficulties arise in even finding the equations of motion, never mind solving them. Material media can respond to applied forces and torques in complicated non-linear ways. Forces can take on forms that make the usual assumptions of continuity and smoothness inapplicable, as in shock waves in fluids, where the very idealization employed makes familiar assumptions about smooth behavior inapplicable at the shock front. Bodies can show "hysteresis," apparent dependence of present response on the past history of the forces to which the system has been subject, not just a response to present applied forces. Friction and dissipation can transform overt momentum and energy into the hidden motion of heat, making some versions of conservation rules hard to apply. The kind of formalization we are now looking at seeks a general framework in which one can construct the appropriate mathematical apparatus for dealing with all of these complex applications of dynamics.

The overall program relies on that generalization of Newton's original approach that invokes forces and torques, the application of the former resulting in changes of linear momentum and the application of the latter resulting in changes of angular momentum, as the primary dynamic elements. The two conservation rules, that of linear and that of angular momentum, are fundamental. Implicitly they rest on the symmetry of dynamics under the translation of systems in space and the rotation of the system's orientation in space. There is also, of course, conservation of energy, corresponding to the symmetry of the dynamics under translation in time, but here the possibility of transformations between overt mechanical energy and the energy of "heat" will ultimately require a dynamics that is integrated with a semi-formalized thermodynamics as well. The account must do justice to the ways in which dynamics transforms from one reference frame to another. Finally it must admit a sufficient apparatus to deal with all the possible subtle ways in which, especially in the theory of continuous media, forces and torques can exert themselves as one component of a system interacts with another.

In one version of this approach W. Noll first introduces the mathematical representative for a body as a piece of a differential manifold isomorphic to a region in Euclidean space. It is given a measure so that the mass density at each point is well defined. Motions are introduced as smooth enough sets of configurations of the body over time that velocity and acceleration are well defined. From the mass density function and the kinematic parameters, momentum and angular momentum can be constructed.

In keeping with the needs of continuum dynamics, two species of forces are introduced: body forces and contact forces. Body forces, gravity for example, are characterized by having a density throughout the body so that the total force can be added up (integrated) from the force density throughout the body. Contact forces, frictional forces for example, are forces characterized such that their application to the body is determined solely by their effect on the boundary of that body. It is demanded of them that they have well-defined surface densities. With this structure in place, one can show that the contact forces can be specified by a function on oriented surfaces of bodies alone.

Next we need the mathematical structure for characterizing dynamics. A dynamical process is a triple of a body, a set of its configurations over time and a family of forces for the body parameterized by the time. The basic axioms of the theory for such processes, the very core of the theory, are the two conservation principles. For any part of a body, the change of linear momentum of that part at a time is equal to the force applied to that part. For any part of a body the change of angular momentum of that part relative to some point of space at a time is equal to the torque (moment of force) about that point applied to that part of the body at that time.

The basic role of these axioms is to allow one to prove the needed generalizations of Newton's Third Law of Motion. Given two separate parts of a body, conservation of linear momentum establishes that the force one part exerts on another is equal and opposite to the force the second part exerts on the first. For contact forces one can show that the force acting across a boundary exterted by a body on one side onto a body on the other is matched by an equal and opposite contact force exterted by the second body on the first.

These results allow one to establish the existence of a basic concept of continuum mechanics – the stress function, a vector-valued function that can be shown to depend only on the normal to a boundary at a point, so that all boundaries sharing the same normal at the point get identical stress-function values at that point, and which is such that the total contact force over a region of a boundary can be found by integrating this stress function over the area of the boundary. With appropriate assumptions about smoothness of the stress function, one can show that the stress function can be given as a tensor operator on the normal vector and, as is fundamental to continuum mechanics, that the tensor so defined is symmetric (which summarizes the results about the mutuality and oppositeness of the interaction forces parts exert upon one another).

What about the issue of how dynamics deals with changing the reference frame relative to which one describes the kinematics of the body in question? Consider any change of reference frame that holds time intervals and distances fixed. Time coordinates can vary only on changing the "zero point" of time. But position coordinates can be changed by translations and rotations of the coordinate frame that are functions of the time. How do the relevant quantities of the dynamics transform with the transformed description of the system? The kinematical quantities of velocity and acceleration will change in their determined way as the coordinate description changes. If we restricted our attention solely to inertial reference frames, the accelerations would remain constant, but we have admitted accelerating and rotating frames as allowed reference frames for characterizing the motion.

What about forces? We want to posit that the "real" forces the parts of the system exert upon one another are independent of the motion of the system. This was assumed implicitly, for example, in Newton's famous Corollaries V and VI to the Laws of Motion. But, if we allow rotating frames, we know that the Newtonian equation of mutual forces and moments of forces with accelerations and angular accelerations won't work. An object at rest in a rotating frame shows motions that require the invocation of additional "inertial forces" as well. These include centrifugal force, the Coriolis force, and an additional term if the axis of rotation is itself not fixed. Noll then provides a definition of equivalent dynamical processes that takes two processes to be equivalent if their respective changes of momentum and angular momentum are related to their respective forces. These forces are related to one another by a transformation that adds the inertial forces to the mutual forces transformed from one system to the other, which are posited as being of fixed magnitude. The inertial force is solely an external body force.

This dynamical framework is useless for predictive purposes until one adds some additional equations that specify the particular constraints imposed on a system and the particular forces and torques the components of a system exert upon one another. These might be in the form of simple constraints, such as demanding rigidity of a body, or they might be in the form of functions specifying how the stresses one part of a system exerts on another vary with the motion of the body. A "principle of objectivity" is imposed, demanding that the constitutive assumptions also must be invariant under changes of the reference frame. Then, if a process is compatible with a constitutive assumption, it will be the case that any process equivalent to this one will be compatible with the same constitutive assumption. It is this posit that fills in the assumption implicitly made by Newton in his two corollaries that the mutual forces (and torques) of interaction in the system are independent of the system's state of motion.

This decomposing of the total theory into a set of general dynamical principles and an unspecified and unlimited number of possible constitutive assumptions provides a nice framework for the theory. It is a framework applicable also to thermodynamics, where one has the distinction between the three fundamental thermodynamic laws (the Zeroth, First and Second Laws of Thermodynamics) and the constitutive descriptions of systems (say as ideal gases or some modification thereof).

The particular framework presented by Noll is meant only to be a very partial account of dynamics. It makes a number of useful assumptions about the continuity of forces, for example, and it treats torques as reducible to moments of forces about a point. But contemporary continuum mechanics introduces devices outside the reach of this framework. Singular forces concentrated at a point, line or surface, contact torques not reducible to moments of forces, "sliding, impact and other discontinuities, singularities and degeneracies" all appear in real continuum mechanics. Even more importantly, the framework does not take into account the melding together of thermodynamic principles and mechanics necessary for taking account of the transformations of mechanical work into heat and vice versa. And that would require for full generality a thermodynamics that can handle characterizations of systems that are not in equilibrium in order to deal with, say, "rates of entropy production" in mechanical fluid processes of viscid media.

The purpose of this "informal formalization" is not philosophical. It is not some "interpretation" of dynamics that is driving the program. The search, rather, is for a sufficiently rigorous presentation of a framework in which one can embed the necessary mathematical tools of the complex branch of applied mechanics that is continuum mechanics. Yet the program does point to a loose philosophical standpoint. This is one that takes as basic the framing of dynamics in terms of irreducible notions of force and torque, and the generalized Newtonian laws of the correlation of accelerations (linear and angular) with imposed forces and torques divided by mass or moment of inertia (as intrinsic properties of matter). The two framework conservation laws, which are reflective of the assumed translational and orientational symmetry of space, then ground, again, a Newtonian program that includes a generalization of Newton's Third Law that is applicable to general cases, not just to point masses interacting by central forces.

Not surprisingly, someone working in this program will usually think of the other possible ways of providing fundamental principles for dynamics – say by the use of extremal principles, or by a dynamically generalized principle of virtual work, or by one of the derived sets of laws such as those of Lagrange or of Hamilton, or by the general methods of Hamilton and Jacobi – as providing useful mathematical reformulations of the basic theory. But, it will be argued, these reformulations are merely derived consequences of the basic laws governing the proportion of acceleration to force and the symmetry of forces induced by conservation principles. Indeed, it will be argued, they are often derived methods whose applicability is limited to special conditions, holonomic constraints, say, or forces that can be derived from a potential, or torques that can be expressed as moments of forces.

On the other hand, there is a long tradition that argues as follows. The appearance of such forces as friction, or of the transformation of energy out of mechanical motion into something else (heat), is just the result of not treating the physics of matter at a deep enough level. Matter is composed of atoms, it is not a continuum as it is idealized in continuum mechanics. The forces binding the components of the atoms together, and, as a consequence binding the atoms together into molecules and the molecules into macroscopic matter, are all forces that can be generated from conservative energy potentials. And all these forces can be dealt with by the more elegant approaches of Lagrangian and Hamiltonian theory more aptly than by using the "Newtonian" program of forces and torques.

Claims and counter-claims about the status of "force" in the foundations of dynamics are made in a number of ongoing disputes. Some of these disputes are most closely connected historically to issues of meaning and ontology. Is Newton's Second Law a definition of force? Is it an analytical statement, which is true by definition? Is it, perhaps, a priori but not analytical, maybe a Kantian synthetic a priori? Or is it just an inductively confirmed empirical hypothesis? Are forces "real," or does the ontology of the world consist only of particles and their motions?

But there is another way in which the status of force in dynamical theory can be controversial. This is an "internal" dispute within the domain of those who apply the theory, and it is more a dispute about which of the many possible basic formulations of dynamics should be taken as "most fundamental" from a physical point of view. Should we start with forces and torques and the conservation laws? Or with a dynamical principle of virtual work? Or, perhaps, with an extremal principle? Or with Lagrangian or Hamiltonian versions of the theory? Or, perhaps, with one of the many ways in which dynamics can be represented in terms of geometries of configuration space or of phase space?

Sometimes these issues can take on a purely "pragmatic" form, the question being one of the ease with which a way of representing dynamics can be used to deal with some specific set of problems to which the theory is being applied. But in other cases it seems that those broader methodological and metaphysical issues are at stake.

The debate in this larger framework also has its interesting historical antecedents. Consider the proposals by Descartes and Leibniz and their followers to frame dynamics entirely in terms of energetic considerations and on conservation principles. Alas, this program couldn't take one much further than Huyghens' brilliant use of conservation of momentum and energy to completely solve the two-body elastic-collision problem. After Newton the theory was dominated by the approach using force and its correlation to acceleration, and then by the generalization of this by Euler to include torques and angular acceleration as well. But the later Lagrangian and Hamiltonian programs showed that at least for a wide variety of cases (holonomic constraints, forces derivable as gradients of potential energy) a subtler "energetic" approach could often provide a far more elegant and "enlightening" approach to a wide range of dynamical problems than could the force approach.

But which approach ought to be taken as "fundamental," as "at the bottom" of dynamics? This is an issue to which we will return when we discuss extremal principles and other non-force methods in Chapter 21. One response to the question, of course, would be to deny its very sense. Why not simply understand that there are multiple versions of the theory, that they are such that for specific application circumstances they can be shown to be mathematically equivalent to one another, and that the choice of "version" is a matter left up to the theorist.

But when theorists deny this "pure pragmatist" response the debate takes on an interesting aspect that we shall encounter again both in our discussion of Machian theories in the next chapter and in Chapter 21.

19.2 "FORMAL" FORMALIZATION: MASS AND FORCE

The aims of the informal formalization we have just looked at are to find appropriate axiomatic formalizations of dynamics that will express the basic components of the theory, namely its general dynamical posits and the framework for the constitutive equations that characterize individual types of dynamical systems, in an elegant manner. This formalization aims to elucidate the appropriate quasi-rigorous mathematical framework in which dynamics can be applied to a range of types of systems, including complex continuum systems.

The aims of the formal formalization we will now look at are rather different. Here the hope is to apply some of the tools of modern formal logic and set theory to answer long-standing conceptual questions about the nature of dynamics. How, for example, do terms such as "mass" and "force" acquire their meanings? Can these terms be defined in other, more kinematic terms such as the positions, velocities and accelerations of the components of a dynamical system? What is the status of the basic laws of dynamics? In particular, what is the status of Newton's Second Law? Is it a priori or a posteriori? Analytic or synthetic? Contingent or necessary?

The current approach to resolving these questions has a number of historical sources that led up to it. In the long tradition of empiricism the logical positivists took primitive meaning to arise out of an association of words with observable elements in experience. Propositions dealing with the unobservable were frequently characterized as lacking genuine meaning at all. But, being "scientifically" minded philosophers, they were reluctant to dismiss all the reference to unobservable features of the world that occurs in scientific theories as meaningless.

Some early suggestions were to take theoretical terms as meaningful only if they could be explicitly defined in terms of the "observation vocabulary." Often it was claimed that some of the propositions of a theory could be taken as fact-stating, but that others were "analytic," since they served as nothing more than definitions of theoretical terms apparently referring to the unobservable in terms of the observation terms. Reichenbach, for example, speaks of "coordinative definitions" as serving this defining function. Some worries about dispositional terms, which could be defined only using modal definitions, which are unappealing to most positivists, led to constructions such as Carnap's "partial definitions" or "reduction sentences."

Later W. Quine and others proposed a more holistic approach to meaning in theories. Here one took the unit of discourse that gave meaning to the theoretical vocabulary as the whole theory. Whereas some of the vocabulary of the theory could still be thought of as acquiring its cognitive role through a kind of direct association with some element of experience – with these elements of experience variously thought of as sense-data, impingments on sensory nerves, or the "observable" ordinary objects and properties of everyday experience – the remaining, "theoretical," terms acquired their meaning by virtue of the role they played in the theory as a whole. From this point of view it would be mistaken to distinguish "analytic," meaning fixing, propositions of the theory from "synthetic," fact-stating propositions. The entire network of assertions played the role of simultaneously giving a cognitive role to the theoretical terms and making factual assertions. The original empiricist theme that only that which had empirical consequences was cognitively legitimate was preserved in the notion that as a whole a theory had a legitimate place in science only if it made predictions regarding the correlations of observable facts with other observable facts.

An early formal version of theoretical holism, long predating Quine, is that of F. Ramsey. Assume that we can divide the descriptive vocabulary of a theory into observation terms and terms referring to non-observable properties and relations. Assume that the theory can be expressed in terms of a finite number of axioms framed in first-order logic. Join the axioms together by "and" signs into a single conjunction. Bracket this conjunction. Replace each of its non-observation predicates and relation symbols by a second-order predicate variable. Preface the new form by existential quantifiers over the new predicate variables. This new theory axiomatized by a single second-order sentence will have all, and only, the consequences framable in observation predicates alone that the original theory had. We can then think of the original theory's use of "positively charged," for example, as just asserting that there is some property which has all and only the features attributed to positive charge in the original theory.

More recent versions of this approach usually move to the language of set theory and offer a "semantic" rather than syntactical reconstruction of theories. One starts with a domain of entities and a class of "observable" relations among them. To impose lawlike constraints on these relations one offers a theory. This consists in the assertion that the observable structure (the domain of observable entities and the observable relations among them) can be embedded in a larger set-theoretic structure characterized by the stated existence of some class of functions and relations and structured posits over their inter-relations.

Much of this work is inspired by results from the theory of measurement. Consider a domain of objects that are related by (observable) "balancing" relations (such as *a* balancing *b* or *c* being out-balanced by *a*). When, and only when, certain regularities hold among the balancing objects (such as transitivity of balancing so that, if *a* balances *b* and *b* balances *c*, then *a* balances *c*), we can assign real numbered "weights" to the objects such that *a* will balance *b* if and only if *a* and *b* "have the same weight," and *a* will out-balance *b* if and only if the weight assigned to *a* is larger than that assigned to *b*. So the facts about the balance relationships among a collection of weights can be represented by embedding the weights-plus-balancing-relationships structure in a set-theoretic structure defined by the assertion that there exists a properly characterized weight function that assigns to each object its numerical weight.

As a simple example of how this can work, consider a domain of elements with two properties, that is selections of subsets from the domain. Call the subsets A and B. Assert that this set-theoretic structure can be embedded into one in which there is a one-to-one onto function, a bijection, from subset A onto subset B. This then constrains the structure to be one in which one has the same number of elements in A as there are in B. One can then use the assertion of the existence of the appropriate function to define an "A equi-numerous with B" structure.

This structuralist view of theories goes a long way toward clarifying a number of long-standing issues. Suppose one of the theoretical terms is a function assigning values to some ordered set of the elements of the observables. For example, consider the mass function in dynamics that assigns a real number to each object, or the gravitational force function that assigns a number to an order triple of two objects and their separation. One can ask, for each possible behavior at the observable level (say each set of trajectories of particles in dynamics), whether that behavior allows one to assign unique values to the theoretical functions. In many cases one can show that such "definability" is impossible. Yet the positing of the existence of the theoretical functions may still constrain the observable behavior. So we can have essential theoretical postulation that does empirical work without the kind of definability of theoretical terms demanded by old-line positivism.

One also gains insight into old questions about the "eliminability" of theoretical terms. Given a structuralist rendering of a theory, can an alternative structuralist account be found that drops one of the theoretical terms but still constrains the possible behaviors at the observational level in the same fashion? Older positivist accounts sometimes dealt with this issue by asking for possible theories that constrained the observables to be framed entirely in the observation language. The formal issues here were of the sort of whether such a purely observational-term theory could be captured in some finitistic way, rather than as a mere infinite list of all the possible observational consequences of the originally, finitistically expressed but theory-term-laden, theory.

Structuralism gives us a clear notion of "Ramsey eliminability," in which one asks whether one can find a thinned-down structuralist account that constrains the observable consequences but that drops one or more of the posited theoretical relations or magnitudes. A standard paradigm of some relation not being Ramsey-eliminable is that of the theory of equinumerosity noted above. Even an infinite list of the consequences of the theory that posits a bijection between the A and B classes will not contain the information that there are just as many elements in A as in B that is contained in the original Ramsey structuralist posit that such a bijection exists between the A and B classes, and no first-order theory about the elements of A and B will do the job either.

These notions can go some way toward illuminating familiar issues about Newtonian dynamics. Mach showed how we could define mass ratios in terms of relative motions for isolated two-particle-collision systems. The mass ratio of the objects is fixed by the antecedent and consequent motions of the particles. But are mass ratios in general definable from the motions of the objects in a system? Earlier results showed that this wasn't so. Even for point particles interacting with each other by two-particle forces, there will be cases where the same relative motions are compatible with more than one set of mass ratios of the particles, for some initial conditions of the system. For another example, consider point masses acting on one another by central forces and obeying the Newtonian laws. In this case the force function is Ramsey-eliminable from the Ramsey specification needed to fix the relative particle motions by means of use of the conservation laws for linear and angular momentum, but, apparently, the mass function is not.

The usual examples of this kind of formalization applied to dynamics use the restricted version of dynamics confined to point particles interacting by means of central forces. Truesdell has strongly objected to this, arguing that it is the much richer theory of continuum mechanics that ought to be the subject of formalizations designed to enlighten us about the nature of dynamics. This isn't merely an insistence that the full complexity of dynamics shows up only in the continuum case, but, rather, that one can easily be misled about important issues if one sticks to the simpler idealization. For example, we know that a principle of conservation of angular momentum must be posited as a fundamental independent conservation rule in Newtonian dynamics. But in the simple theory of point-particle mechanics with central forces, the conservation of angular momentum follows from the conservation of linear momentum, as Newton showed.

Or, again, consider the essential role played in the theory by the force concept. The fact that force is Ramsey-eliminable from the formalized theory of point particles interacting under central forces might make it seem an extraneous notion that can be dispensed with in favor of the conservation principles. But in full continuum mechanics, with its body forces and contact forces, there is little hope that force could be Ramsey-eliminable in the more complicated theory were it ever formalized sufficiently to be given the full Ramsey formalized treatment. Restricting our attention, then, to the formalization of the pared-down models of point particles and central forces can mislead us if we are using the formalization to try to decide what is essential to the full dynamical framework.

This approach to the formalization of physical theories has some virtues that are rarely questioned. Consider the old controversy over whether to view the Newtonian Second Law as an "analytic" definition of "force," or instead as an empirically grounded postulate of a "synthetic" nature associating forces with accelerations. To get any kind of Ramsey-formalized Newtonian theory to generate specific observable relative motions for some system of objects, one must plug into the framework both the general Newtonian postulates and also constitutive equations that relate forces to the state of the system. So "force," or, rather, the second-order predicate variable that will take the place of that term in the Ramsified version of the theory, will appear both in the Second-Law position and in the constitutive-law positions. Take the Ramsified theory as a whole as both constraining the observables, that is, making empirical predictions, and specifying the meaning of any "place-holding" predicates at the non-observable level such as "force." There is no further need to discriminate between components of the Ramsey sentence (or set-theoretic definition) and to declare some of them analytic and others synthetic. It is the Ramsified theory as a whole that simultaneously imposes constraints on the observables and establishes the structure that gives the place-holders their "meaning" in the original, un-Ramsified, version of the theory.

The whole project of understanding theories from this Ramsey point of view requires the sorting out from the collection of properties, relations and functions spoken of in the theory those that are to count as the "observables." These need not be thought of as being observable features in some ultimate, deep, philosophical sense. But they are to be understood as being represented by concepts whose meaning we grasp independently of the theory being formalized. Then the remaining "theoretical" concepts are understood as having meaning only insofar as they serve as place-holders in the theory to be eliminated in favor of the second-order variables.

But what are the appropriate observables for a formalization of Newtonian dynamics? The usual approach is to take these to be the relative positions of the material objects in the system and their changes over time. But what kind of "position" is needed? If you are doing standard Newtonian dynamics you need positions specified relative to an inertial reference frame. This could be added to the usual versions of Ramsified dynamics, perhaps, by taking the notion of some rigid reference frame fixed by distance-measuring devices taken as primitive and from outside the theory, and then adding existential quantification to the formal theory to the effect that there is some such frame such that relative to it positions as specified will have their changes over time embedded into the framework that has taken the place of the usual dynamical laws and constitutive force equations.

But this is not the only version of Newtonian dynamics in existence. A Machian approach will require working in a framework of relative positions taken as specifiable from outside the theory by means of distance-measuring devices taken as primitive, and then finding the inertial frames from within the theory in terms fixed by the dynamics of the entire Universe.

Issues of time need careful thought as well. The existing versions of formal formalization of dynamics take the time function as given from outside the theory by means of some accepted "clocks." But the issue of the source of the appropriate time measure, in other words the response to Newton's insistence on a notion of absolute time, is often dealt with by arguing that the "natural" measure of time lapse comes from inside dynamics itself. This is the Machian approach that takes the "ephemeris time" of the universal system as giving the "absolute" time scale that will work appropriately to serve as the time measure for any temporarily isolated system within the cosmos evolving according to Newtonian dynamics. All of this will be discussed in the next chapter.

But if this approach is followed, a careful rethinking of how to formulate the "from outside the theory" concepts dealing with spatial position and temporal interval, and hence how to treat the "from outside the theory" notions of relative velocity and acceleration, which respects the problematic issues of choosing an appropriate spatial reference frame and an appropriate time measure, will be needed. What to count as the concepts given independently of the theory and what to count as the concepts whose meaning arises solely by virtue of the place-holding role played by the concepts in the theory is by no means obvious.

Here we face a general issue with this Ramsey approach to formalizing theories. If one were using a Ramsey formalization from the perspective of some radical positivism that made a once-and-for-all distinction between the genuinely observable terms of a theory (whether these are referring to sense-data or to ordinary-sized physical objects and their features or whatever) and the genuinely non-observable terms in the theory, the issue of which terms to leave as they are in the theory and which to replace by second-order variables and quantifiers over them would be settled. But, in the absence of such a philosophical stance, the issue of which terms to count as those referring to the "observables" becomes problematic. We do intuitively feel that some features of a mechanical system can be determined by us totally independently of our dynamical theory and ought to count as observables. Whether two particles are "coincident" in space at a time might be such a feature of the world. But other features may remain problematic in their status: Can they be determined independently of the dynamics, or are they, rather, as mass and force are always taken to be, fixed by the theory and determinable only within the theory?

So, while we may want to take the spatial separation between two particles as determinable by some standard means for measuring spatial distance (rigid rods?) as being an "outside the theory" notion when formalizing dynamics, the need to refer accelerations only to genuine inertial frames if we are to get the right observational predictions out of the theory warns us not to take "position" itself as an observationally, dynamically independent notion. Similarly, although we may want to think of temporal order of events as something open to us without any dynamical presupposition, the notion of a good clock, that is, of the appropriate manner of assigning metric temporal intervals to pairs of events, may be available only within the dynamical theory itself.

We must be grateful for the Ramsey-style holistic account of theories for letting us resolve, or rather dissolve, the old contention between those who would take the Newton's Second Law as analytic since it is definitive of "force," and those who would call it synthetic. Let us agree that meaning attribution and fact-determining value are both spread throughout the theory taken as a whole. Another way of putting this is to assert that it is theories that are the units of scientific meaning – not words and not individual sentences of the theory, as was commonly believed prior to "Ramsification."

But what about the old "metaphysical" debates about force? Is force some real element of the world, as much a part of our ontology as the things and properties of the observable realm, or is talking about force nothing but a way of representing the regularities among the observables?

First look at the representational side of the debate from the Ramsey point of view. We have the observable things and their properties and relations. Then we have the embedding of the structures among the observables into the larger structure asserted by the Ramsey sentence for the theory. But all the Ramsey sentence says is that there exist theoretical relations that have whatever structural relations to the observable that the original theoretical relations had to those observables in the original, pre-Ramsified, version of the theory. But as far as ontology is concerned, as far as commitment to being is concerned, we might as well take the Ramsified version of the theory as just telling us that the observable structure of things can be embedded in an abstract structure of numbers, sets and the like. So, to take the example from the theory of measurement noted earlier, all the assignment of weights to things does is keep track of the facts about balancing of entities and the concatenations of entities by assigning appropriate numbers to objects, namely their weights.

Here is an example from particle physics that makes this representationalist reading of the theoretical terms seem plausible. A new kind of particle is discovered. It is produced quickly in oppositely charged pairs. But the individual particles, once produced, decay slowly. But the basic theory predicts that any reaction and its opposite will have the same time scale. The solution is as follows. Invent a new theoretical property, "strangeness." Attribute +1 unit of strangeness to one particle of the duo and −1 unit of strangeness to the other. Invent a new law – that the total amount of strangeness in the world cannot change in a fast process but can change in a slow process. The pairs can then be created quickly since the total strangeness change is zero. But the individual particles can decay only slowly since their individual decays change the total strangeness in the world. It is more complicated than that, but the point still holds. Is there any reason to think that talking about strangeness is anything more than keeping track of the ways in which these particles are created and destroyed by assigning appropriate numbers to them and imposing a rule on these numbers that forces the observed regularity among what we are now taking as the observables?

So we start with our theory that posits a strangeness function that assigns numbers to certain particles and accounts for the behavior of those particles by embedding strangeness into some set of laws. Then we Ramsify the theory, ending up by just saying that "there is a function such that . . . ," where the ellipsis is filled in with everything the original theory said about the strangeness function. Do we then worry about such things as whether "strangeness is real" or "what the nature of the strangeness property is"? Or do we, rather, simply read our theory as embedding the particle behavior into an abstract structure that constrains the lawlike behavior of the particles, with strangeness as nothing more than a device for assigning "tracking" numbers to particles? And, just as it is for strangeness, isn't it so also for electric charge and the like?

For a contrasting case, consider the way we treat "unobservable" spatial dimensions of things when we talk about, say, the relative sizes of atoms and their nuclei. Even if such spatial dimensions can be ascertained only indirectly through the consequences they engender among the observables according to some theory, don't we think of these unobservable physical dimensions as just as real as observable spatial dimensions? And, even if we do Ramsify our theory, don't we think that "we knew what properties we were talking about" when we spoke of those hidden spatial dimensions in our pre-Ramsified theory?

The issues here form a complex of some of the most intractable problems in philosophy. The basic problems of meaning of concepts in our theories, and of the commitment we make when we accept our scientific theories, remain among the most difficult unresolved general problems in methodological philosophy of science. How do the terms of our theories acquire their meaning? What kinds of answers can we give to the question "What is the nature of that meaning?"? How is the nature of that meaning related to the issue of how we are to understand our ontological commitments that come along with our acceptance of a theory? And how are all of these issues related to a presupposed distinction between the observables and the unobservables spoken of by the theory?

There are those who will deny any principled distinction between the concepts of a theory that allegedly refer to observable entities, properties and relations and those that allegedly refer to the unobservables. It is a familiar argument in philosophy of science to claim that all properties are in principle observable properties. This argument starts with a rejection of an old positivist line that the observables are properties of "ideas in the mind," rejecting such a claim as based on false theories of mind and claiming that such a proposal would block us forever from a comprehensible account of an objective world.

But if we take the observables to be some subset of the objective physical world, then a simple slippery-slope argument takes us from the intermediate-size objects and their properties of everyday perception, down through those features available to us by means of simple optical instruments, to the claim that, since all posited physical features are available to us through the empirical results of some instrumental device, all those features ought to count as observable features of the world.

But it is just such theories as ones like the dynamics we have been exploring that convince others that a hard-and-fast, principled, distinction between the observable and the unobservable must play a crucial role in our understanding of our foundational physical theories. In dynamics it is spatial relations among material objects and temporal relations among material events that seem to be taken as the observable, with mass and force as the unobservable. In discussions of the geometry of spacetime theories it is often coincidences of events picked out by material objects (intersections of paths of particles or of light rays) and coincidences of events on spatially coincident clocks (local simultaneity) that play the part of exhausting the observable, with the geometry of "spacetime itself" as the unobservable. In interpretation of quantum mechanics the results of outcomes on measuring devices are taken as observable, with the quantum state function itself taken as dealing with something that is unobservable in principle.

Suppose we accept such a principled observable/non-observable distinction as legitimate. Still we must deal with issues concerning how that distinction functions crucially in some theory of the attribution of meaning to terms in theories. The standard line has a long history in philosophy. It ties our mode of epistemic access to features of the world to an account of how our concepts referring to those features acquire their meaning. The standard line goes like this: For features of the world accessible to our observation, concepts referring to those features acquire their meaning by the process of ostensivity. We learn the meanings of the terms by some process of direct attachment of term to feature, where the feature is presented to us in our observational experience. (Mother points to a red object and says "red." Child learns to designate redness by "red.")

Now the legitimacy of such a theory of meaning and its accrual has been subjected to savage criticism, most notably by Wittgenstein in his *Philosophical Investigations.* Here the fundamental problem is that of how we learn to "go on" by this alleged ostensive process. What allows us to know that it is redness that is intended in the ostensive definition, not one of the infinite number of other features present in the experience? But let us put that to the side and assume that something prior to and outside of a theory itself provides us with a grasp of the meaning of the observational terms present in the presentation of the theory.

But then the question remains how meaning accrues, and what kind of meaning accrues, to those terms of the theory that do not refer to observable features of the world. The canonical answer, of course, is that of Ramsey–Quine theoretical holism: the non-observational terms acquire their meaning by virtue of the role they play in the theory. And the full story of their meaning is given once that "role in the theory" is pointed out. Or, in formal terms, the Ramsey sentence for the theory does all the work the original theory does and so explicates all there is to say about the meanings of the original theoretical terms.

Suppose we accept that view about the meaning of the non-observational terms. What consequences should that have for our views about the ontology of our theory? In our particular case, what consequences should this view about meaning have for the old questions about the "reality" of mass and force?

The first answer that will come to mind is simple representationalism. The theory simply asserts that the lawlike structure governing the real entities and features of the world, those general features of the *observable* things and their properties, can be embedded in an abstract structure that is characterized by the Ramsified version of the theory as a whole. So talk about the unobservable, in our case about mass and force, is just talk about how numbers can be assigned to objects or pairs of objects that keep track of how their relationally characterized spatial features vary with time. There are particles, for example, and there are their spatial separations. And there is time and there is a variation of these spatial separations in time that depends upon the specifics governing the interactions of the particles. These features of the world can be tracked by assigning numbers to the particles (their "masses") and numbers to pairs of particles (their interacting "forces"), but there is nothing to mass and force beyond such tracking numbers. And if multiple, distinct, representational embeddings of the laws governing the observables will do the trick, then each such representation is a "true" characterization of all that is real about the world.

Why is such pure representationalism unpopular with many? Well, it leads to "irrealism" with regard to all of those entities, properties and relations posited by our theories that are in principle unobservable, and to which reference can be made only by terms whose meaning is completely specified by the place-holding role these terms have in the theory. But can we really dismiss from our posited ontology of the world all that is unobservable? Taking the limits of that with which we can be observationally acquainted as the limits of what there is seems too extreme a positivism for many. And if we get pushed onto the slippery slope that sometimes leads philosophers to take as the observable only the immediate, subjective contents of perception, then representationalism will lead us to a quasi-"Pythagorean" world in which beyond the contents of minds there is nothing but numbers!

One realistic approach will stick to the Ramsey idea that the meaning of such theoretical terms as "mass" and "force" is fully exhausted by their roles as place-holders in a theory. But it will deny the representationalist ontology that some take as following from such an account of meaning.

Posit that the theory in its Ramsey formalization should be taken as asserting not only that properties exist that fit the formalism's constraints, but also that a unique such property is being posited. Next admit that more than one property can exist that fits the theory's formal constraints. But insist that the new, modified, Ramsay sentence is claiming that there is a unique property that fits the formal demands, but is the property that in our world really explains (along with the other properties posited by the theory) why the observables behave the way they do. That observable behavior is the one codified by the observational consequences of the theory in its new, modified, Ramsey form.

But what is this property? How can we come to know what it is, above and beyond the characteristics it is constrained to have by the Ramsey formalization of the theory? How can we say, for example, which property of the many that fit the formal characteristics is the unique one that is really doing the explanatory work in our world? The answer is that there is little that we can say to answer these questions. We can specify the specific property in question by asserting that the property in question is just "the one property that has the formal characteristics that are actually present in our world and that actually explains (along with the rest of the theory and the rest of the theory's ontology) the observable consequences of the theory." But that, of course, doesn't answer our nagging longing to know "what that property is" in the sense that we think we have a full grasp of the observational properties, perhaps by ostensive acquaintance. ("Do you want to know what redness is? Look at this!" – Showing the questioner a red object.)

So, on this account of ontological realism and a Ramsey-sentence version of the limits of our grasp of the meaning of theoretical terms, the world posited by our theories is real but, to the degree that we reach beyond the observable, ineffable. Here historical memories of Kant's "noumenal world" or "things in themselves" may occur to the reader. And familiar empiricist doubts will arise about the legitimacy of positing "somethings I know not what" as explanations. But if we once accept the legitimacy of the claim that some terms in our theories, if they are construed as having genuine reference in the world rather than as playing a merely representational role, can only refer to features of the world that are forever immune to being comprehended by observation and experience, then this kind of understanding of our Ramsified theories may be unavoidable.

Another approach makes a distinction between two kinds of features said to exist by the Ramsey sentence of a theory. One set of features is taken to be those whose nature is the same as that of features available to us by observation and spoken about in the observational consequences of the theory. The other posits are of features not so knowable by "analogy" with observational features. For the latter posits a representationalist account is given. But for the former it is alleged that the posited features can be taken as "the same properties in the world" as their analogical counterparts in the observational assertions.

It is one thing to replace "strangeness" (as used by particle physicists) with a Ramsey variable and a quantifier. We have no further understanding of strangeness than the role it plays in the theory, and we may just as well construe reference to it as doing nothing but representationally keeping track of how things behave. But we talk of the location of particles "too small to see," and of their size. And we talk of the time interval between events outside the range of our perception. And we think we mean the same thing by the spatial and temporal terms even if, being outside the range of our perception, they must appear in the Ramsey version of a theory in the form of quantified variables only.

Perhaps this explains the frustratingly ambiguous attitude that has been taken toward "force" ever since its introduction into dynamical theory. For d'Alembert it is a useless concept best excluded from theory altogether. For Boscovich it is the fundamental reality. For Mach it is to be understood by a positivist redaction. For Ramsey theorists it is a quantified variable.

But we continually feel that forces are "real" and that "we know what they are." Why? Well push on an object and one "feels" the force being exerted. And if the reply comes back that one must not confuse the subjective, kinesthetic muscular sensation we feel with "what is in the objective world," one is driven to the response that our direct awareness of spatial and temporal aspects of the world is equally "subjective." We don't know whether to classify "force" with "strangeness" (easily eliminated except as a representational device) or with "size" (too analogous to observable size to be thought of as just a representational element of the theory).

To get a feel for this version of "partial theoretical realism" consider our understanding of other's mental life. A brick drops on someone else's foot and they yowl "in pain." How can one doubt (despite Wittgenstein's famous mockery) that we know "what they are feeling" from "our own case." It has often been argued that we know what is going on in someone else's mind by "analogy" from our own case (our directly knowable mental processes). But here the claim is that our very understanding of what we are talking about when we talk about the contents of "other minds" is a grasp of *meaning* by analogy with the meaning of terms we understand "from our own case." It is "semantic analogy" that is being posited.

And now it is posited that something of that sort goes on for some of our theoretical vocabulary – we know its meaning by semantic analogy from the meaning of the terms as they are grasped from their application to observable features of the world. It is then argued that only when we have such a grasp of the meaning of the theoretical terms can we take the use of them to be in positing a real ontology for our theories. All the rest of the theoretical talk is mere representationalism.

But such an approach to a "partial realism" for the theoretical part of a theory meets several important objections. First there is the claim that the theory of meaning accrual posited is fallacious in two ways. It is a mistake, it is alleged, to think that meaning accrues to any terms by some ostensive learning that associates the meaning of a term with some observed feature. Next it is argued that it is a mistake to think that one can transfer the meaning of a term in one context (in our case the context of observational consequences of the theory) to its meaning in a radically different context (in our case the role of the term in referring to non-observational features posited by the theory). Worse yet, it is argued, is any claim to the effect that some such meaning accrued by semantic analogy gives us any reason to take a more realistic approach to the ontology of the apparent reference of the term than we do to terms whose meaning is given solely by the role they play in the theory not supplemented by any meaning transference by semantic analogy.

At this point, frustrated by the alternatives of a realism that has our science positing totally ineffable features of the world or a partial realism backed up by the dubious program of semantic analogy, one is tempted to consider the virtues of a full-fledged representationalism. What would be so bad about simply taking terms such as "force" and "mass" as simply placeholders in a theory whose interpretation allows these places to be filled by abstracta such as numbers or sets? Probably the gravest objection to this is a slippery-slope argument. What, after all, are the real observables? Are they not subjective contents of direct apprehension? Shouldn't we, then, treat even our reference to everyday physical objects, such as tables and chairs, rocks and bodies, as theoretical postulates in a kind of implicit, everyday theoretical structure of "common sense." Adopting that view along with a pure representationalism leads us, of course, to a philosophical phenomenalism about the "real" supplemented by a Pythagoreanism (all things are numbers) even about the everyday objects of common-sense realism. If force and mass are nothing but abstracta in a representationalist picture, why aren't hands, feet, clouds and houses?

These issues throw us into the most intractable questions of metaphysics, epistemology and semantics that continue to frustrate philosophical understanding. Perhaps all we can say here is this: The "holistic" Ramsey approach to the understanding of dynamical theory clarifies what we mean when we say that a term like "force" gets its meaning from the role it plays in the total theory. The old debates about whether the Second Law is analytic or synthetic, whether it defines force or is a contingent assertion of the relation of force to mass and acceleration, become less of a conundrum. "Force" is simultaneously defined by and gives empirical content to the Second Law and the constitutive force laws in which it appears. But a Ramsey-sentence understanding of the theory goes much less far in helping us to deal with the long-standing issues of realism versus representationalism in our understanding of theories that speak of entities and features we take to be "beyond the range of our observation."

SUGGESTED READING

A clear exposition of contemporary "informal" formalizations of dynamics can be found in Noll (1974). Sneed (1971) provides a clear exposition of the Ramsey-sentence formalizations of theories along with explanations of such notions as Ramsey eliminability and the limits of definability of theoretical terms by observational terms. There is also a worked study of the application of the theoretical material to Newtonian particle dynamics. For a thorough and deep discussion of the possibility of accepting an ontology of "ineffable" features of the world, see Lewis (2009). For the idea that theoretical concepts could be understood by analogy to observational concepts (and that this might have a bearing on our understanding of the ontology of theories), see Putnam (1962) and Sklar (1980). For Wittgenstein's famous doubts both about the ostensive acquaintance theory of meaning and about the legitimacy of meaning by semantic analogy, see Wittgenstein (1958).

CHAPTER 20

Relationist dynamics

20.1 MACHIAN DYNAMICS

We have seen how Newton attacked spatio-temporal relationism not through a typical philosophical argument that might be argued by a substantivalist, but by claiming that "absolute" space and time were necessary for dynamics. Inertially moving systems are radically distinguished from those which, under the influence of external forces, are not moving inertially. But, to make sense of motion in a constant direction and at a constant speed, we need an absolute frame relative to which constant direction is constant and relative to which our changing positions are to be measured, and we need an absolute lapse of time (up to a linear transformation) relative to which constant speed is constant. The discovery that absolute place and velocity play no role and can be eliminated even in the framework of Newtonian dynamics in favor of Galilean or neo-Newtonian spacetime with its class of inertial frames should give the relationist no comfort.

In Chapter 16 we briefly sketched the version of relationism proposed by Mach. The core of this program sought a "cosmic" solution to the problem of the reference frame relative to which accelerations were absolute, with the source of the inertial forces sought in acceleration-dependent relations of test systems to the distant "fixed stars." We noted that Mach's proposal remained shrouded in vagueness and that subsequent proposals in his vein were generally deemed unsatisfactory. In the twentieth century a new kind of Machianism has been proposed by J. Barbour and B. Bertotti. It is to their theory that we must now turn our attention.

Barbour and Bertotti leave our notion of relative spatial distance intact, taking it from outside the dynamical theory, although Barbour has considered refinements of the theory that seek at least a partial dynamical account of spatial distance as well. They also, unlike a number of Machian programs, take mass as an intrinsic feature of material things and do not seek some dynamical origin of mass itself. On the other hand, the standard of time lapse declared by Newton to require an absolute time is given a different, dynamical perspective by Barbour and Bertotti. And, like Mach, but in a quite distinctive manner, they seek a cosmic origin for the reference frame relative to which accelerations are absolute.

Once again the great H. Poincaré provides an important inspiration for the Barbour–Bertotti program. Suppose we have a system of point particles that are moving under the influence of their mutual gravitational, and perhaps other, attractions and repulsions. Assume the existence of Newtonian absolute simultaneity and methods from outside dynamics that let us determine the definite spatial separations of the particles at any time. Suppose we fix these spatial separations and the mutual relative velocities of the particles to one another at an initial time. Can we then determine, using the Newtonian laws of dynamics, what the relative spatial separations and velocities will be at any future time? We can if the system as a whole is in inertial motion (Newton's Corollary V) and even if it is uniformly linearly accelerated in all its parts (Newton's Corollary VI). At least that will be so if we assume that the interactive forces are frame-invariant in the Noll–Truesdell sense that is also noted by Barbour as being of fundamental importance.

But this will not be so if the system as a whole is in absolute rotation. If that is so, information above and beyond the initial relative separations and velocities is needed if we are to predict these relative quantities at later times. Basically we must know the direction of the axis of rotation of the system and we must know the magnitude of the absolute rotation about that axis. If our only initial conditions for the particles are the relationistically acceptable quantities, then we must somehow encode these system-as-a-whole absolute rotational facts into the dynamics in order to be able to predict even the relationistically acceptable quantities at later times.

A technical device that proves useful for developing this new Machian version of dynamics comes from Jacobi. Frame the dynamical problem for a system as if time were a dynamical variable rather than a parameter relative to which dynamical change takes place. Introduce a new parameter to characterize position along paths in the new extended configuration space that includes the usual Newtonian spatial configurations (specified in an inertial frame, of course) and time as well. Look for extremal principles to develop dynamics in this framework. It turns out that one needs first to specify the fixed energy of a system and then to pick an extremal principle appropriate to that energy. This extremal principle will fix a class of paths in the spatial configuration space that one gets by projecting down from the full space and time configuration space. The paths are distinguished from one another by the different initial conditions holding for each path. Now time is reintroduced into this spatial configuration space by specifying how far along one of the paths parameterized by the arbitrary parameter one has moved in any specified, elapsed, Newtonian, absolute time. Newton's absolute time reappears as a preferred parameterization within this configuration space along the trajectories. So far there is nothing relationistic in this picture, since the spatial configuration space is one that is adapted to an inertial frame.

But now use Mach's suggestion that the secret to the solution of the problem lies in having a cosmic perspective. Stop paying attention to the usual, isolated, small systems treated in dynamics. Instead focus on the dynamics of the entire material cosmos. Assume this cosmos to be finite in extent. Next posit that the material cosmos as a whole has no absolute rotation. This cosmos for us is, of course, far vaster than Mach's "fixed stars." Those were part of our galaxy, which we now know is in absolute rotation. We now have a cosmic model of vast numbers of galaxies. Is there any reason to think that this vast, material cosmic array really has no absolute rotation? The answer is positive. The remnant black-body radiation, which was released into the cosmic vacuum when the expanding Universe first became electrically neutral as the existing charged particles joined one another to form neutral atoms and hence became transparent to free radiation, provides us with much information about the structure of the cosmos. It is probed specifically for indications of overall spatial uniformity modified by the density fluctuations which are so important to cosmology. But it also, allegedly, informs us that the cosmic matter had very little if any absolute rotation at the time of release of this radiation. Such rotation would show up in a spiral temperature distribution for the radiation, which is not observed. So let us take it that the material cosmos is not, as a whole, in rotation.

Next posit the energy of the cosmic system to be fixed (set at zero for convenience). We can then use the Jacobi method to plot a trajectory for the cosmic system through configuration space. But, using the observation of Poincaré, we see that, given that the rotation rate – the angular momentum – of the cosmic system is zero, we need only use a relationistically acceptable *relative* configuration space of relative positions to characterize this trajectory. We do not need the full Newtonian configuration space with its preferred inertial frame coming from outside the system. Each trajectory for this system will be fixed by the initial *relative* positions and velocities of all the components of the cosmic system.

Now, if we look at any "isolated" sub-system of the cosmic system, isolated in the usual sense of not interacting by forces with its surroundings, we can choose for the non-rotating framework in which to describe its motion, even if the isolated fragment system is itself rotating, a rest frame that is not rotating with respect to the cosmic material frame, since we know that that global material frame is not in rotation. So we have replaced Mach's "fixed stars," as the material frame relative to which accelerations are absolute, with the cosmic non-rotating frame. But is it clear that we have made much progress toward relationism at this stage? After all, aren't we still relying on standard Newtonian dynamics with just the "happenstance" that the cosmic material frame will do for absolute space because it happens not to be in rotation?

And what about Newton's absolute time? Here Barbour and Bertotti use a familiar idea from nineteenth-century astronomy. If one wishes to do celestial mechanics, the comprehensive theoretical treatment of the solar system including careful accounting for the complex perturbations introduced by the mutual interactions of all the planets, one needs a very accurate standard of time. Even the rotation of the Earth is not adequate for this purpose, since it is not uniform enough in Newtonian absolute time. The trick is to use "ephemeris time." One builds a theoretical model that takes into account the observed motions of the celestial bodies relative to the "fixed stars" and their mutual gravitational interactions, using the best series approximations and calculations that could be handled at the time. Then one looks for a scaling of the time that best fits all of the data into the Newtonian model one has constructed. That assignment of time to events is ephemeris time.

In the Barbour–Bertotti scheme one simply takes Newtonian absolute time to be the ephemeris time for the cosmic, non-rotating, system as a whole. This is unique only up to a linear transformation, of course, since even for absolute time the zero point and scale of unit time are arbitrary. This cosmic ephemeris time will serve as an appropriate Newtonian absolute time for isolated sub-systems of the cosmos. Once again, this will be true even if those systems are in absolute rotation. In fact it can be shown that if we construct the ephemeris time for one of these isolated systems, even if that system is in absolute rotation, that small-scale ephemeris time will be (to within a linear transformation) identical with the cosmic ephemeris time. Of course, if we are calculating the ephemeris time for one of these isolated fragments of the cosmos, we will have to do so by working in an appropriate "inertial frame."

So far, it can be argued, all Barbour and Bertotti have done is to extract some results of Newtonian dynamics that apply in the special case where an isolated system has no angular momentum as a whole. Then, if the entire cosmic material system has no rotation, these conditions will apply to it. One will be able to frame its initial-value problem in terms of relative positions and velocities only, and track its evolution in relative configuration space. By posit, this cosmic material frame will provide us with a non-rotating frame for its isolated fragment systems, and its ephemeris time will give a scale for absolute time.

The next step, though, gives the Barbour–Bertotti account a more original flavor. Instead of working in a single spatial framework over time, refer each momentary configuration of the system to its own rigid spatial framework. Then construct a new extremal principle as the foundation of dynamics. This will be a least-action principle where the action is calculated by referring the positions and velocities of the system at each moment to its own spatial frame. But demand that the least action be calculated relative to any kind of rigid motion from frame at one moment to frame at later moment. The demand of least action will then, by itself, guarantee that the system has zero angular momentum, since the specification of the action will now have a rotation term appearing in it connecting the frames to one another, and the minimization demand will force zero angular momentum on the system.

Of course, one will only be able to legitimately apply this least-action principle to the cosmic system, since ordinary, temporarily isolated, subsystems of the cosmos will, in general, *not* have zero angular momentum. But, as we have noted, the application of ordinary Newtonian dynamics, which is done by referring these isolated systems to the preferred frames picked out by the cosmic non-rotating frame and using as Newtonian absolute time the ephemeris time of the cosmos, will follow.

There are many other aspects to the Barbour–Bertotti program, including extending the theory to allow arbitrary scaling of the spatial metric framework, a feature simply taken as a given in the narrow version of the theory we have been looking at. And one can extend the program to construct a Machian version of general relativity. Furthermore, Barbour has offered even grander theories extending his way of thinking about foundational theories to the many mysteries of quantum mechanics. But for our purposes the theory sketched above, limited to a Machian relationist account of Newtonian particle mechanics, will do. As we shall see, this program gives rise to a number of fundamental methodological issues: about the sources of the demand for relationism; about the ways available to reconcile relationism with the Newtonian dynamical facts; about the role cosmic concerns may play in grounding a theory that seems initially to be about local systems; and about what counts as a good reason for accepting or rejecting a foundational theory in general.

20.2 PHILOSOPHICAL QUESTIONS

Whether or not one accepts a Barbour–Bertotti approach to a Machian reformulation of Newtonian dynamics, an examination of the schematic structure of their theory provides an ideal context for discussing a number of philosophical issues that have their interest both within a discussion of the foundations of dynamics and beyond that context.

Approaches to dynamics that follow the route initially marked out by Mach have as at least part of their motivation the general theme of relationism. Following Descartes, Leibniz and Huyghens, and in contrast to the firm substantivalism of Newton, they argue that we should formulate our theories in such a way as to invoke only the relationally acceptable spatial and temporal concepts. No toleration of a substantival space, or even of such weakened substantival concepts as Galilean spacetime, should be allowed. Nor should we take as a given any "absolute" measure of the lapse of time taken from outside of dynamics itself.

Now some reasons for rejecting substantivalism are those that Leibniz adduces in his correspondence with Clarke. Here the problems of a non-relational account are largely traceable to the idea that a substantival space and time is built of qualitatively identical spatial locations, and has qualitatively alike spatial directions and qualitatively identical temporal moments. So all Leibniz's arguments about indiscernible places of matter in space and motions of matter relative to space come into play. But Leibniz is stymied by Newton's argument to the effect that some kinds of motion relative to space itself, accelerations including rotations, do show discernable effects and that only one measure of the lapse of time will correctly scale to give the simple laws of dynamics.

But we can satisfy some of Newton's needs and at the same time take the sting out of some of Leibniz's arguments without the need for a sophisticated Machian approach. Allow oneself the supplementation of the usual relationist spatial and temporal relations with additional monadic predicates applicable to dynamical systems. These can include "magnitude of absolute acceleration" and "magnitude of absolute angular momentum." An object accelerated with respect to space itself, for Newton, and with respect to the inertial frames, for a neo-Newtonian, is now thought of as just having a monadic quantity of absolute acceleration rather than the magnitude of some relational feature with respect to space or the inertial frames of spacetime. Other than this change in the way of thinking about absolute acceleration and rotation, Newton's theory, or its filled-out version constructed by Euler, is left alone.

This is, of course, a very "cheap" way of obtaining "relationism." Indeed, there are those who wouldn't think of it as relationism at all. After all, there still is a very real distinction between systems that are in absolute acceleration and those that are not. But it saves the Newtonian need for the observable dynamical distinctions, and it abandons the "real spatial locations" or "real spacetime structures" relative to which the absolute motions are motions in the older substantivalist versions of the theory. Indeed, there are even passages in Newton and Leibniz that suggest that they might have been thinking along these lines.

Now one objection to doing things this way is that this theory "fails to explain" features of the world the Newtonian version explains. Two buckets of water are presented. They are such that one spins relative to the other (and vice versa). But in one the inertial features show up in the curvature of the surface of the water, in the other they do not. Why? Newton has an answer: One body of water is spinning relative to space itself. In this new account what do we say? Well, that one body of water has the internal quality of real spinning and the other doesn't. But is that explanatory at all? Isn't "having the property of real spinning" nothing more than showing the inertial effects?

If you like. One could properly say that this new account distinguishes the special class of moving objects that show no absolute acceleration or rotation and offers no further explanation as to why these are special. But one could also say that the Newtonian "explanation" of the facts about these systems, namely that they are in uniform motion and are not rotating relative to space itself, is a very thin explanation. For the only way we have of identifying these "relative" motions to space itself is through the appearance of the inertial effects. What is empirical on both accounts is the relative change of inertial effect with respect to relative motion and the fact that some systems are, indeed, devoid of any inertial effects.

But the Machian, it seems, wants more than this "cheap" way out for the relationist who holds to Newtonian dynamics. To what extent does this new Machianism provide us with a deeper kind of relationism?

First look at time. From an orthodox Newtonian standpoint one could already think of "absolute time" differently from the way Newton did. Instead of talking about some "time itself" "flowing equably" and the like, just take Newton as having shown us that there is a profoundly preferred way of scaling time intervals. Take any isolated system. Refer its motion to an inertial frame. There will then be a way of scaling time that radically simplifies its dynamical equations – the scaling that is one of the preferred Newtonian measures of lapse of time. It is a very deep fact about the world, then, that the preferred time scalings for each of the isolated systems are all linearly related to one another. Absolute time for one system is absolute time for any other.

The Barbour–Bertotti approach is subtly different from this way of reading the notion of absolute time. Here one begins by finding the preferred time scaling for the entire cosmos, the "ephemeris time" of the Universe. Only after this is it then shown that this time scale will provide the usual Newtonian preferred time scaling for describing the motions of isolated fragments of the cosmic system. But, just as in the previous interpretation, the "absoluteness" of absolute time is revealed as the simplifying time parameter for characterizing the motions under the Newtonian dynamics.

What about absolute space, or, rather, the absolute inertial frames? If we stayed with the version of the Barbour–Bertotti approach that simply notes that a Jacobi-like variational principle in full configuration space can be replaced by such a principle in relative configuration space when a system has no spin, it would seem implausible to think that much progress has been made toward a relationist theory. Introducing the cosmological elements takes us a little further. Here the fact that the cosmos has no rotation (if that is a fact) plays a role quite similar to that played in Mach's original work by the important empirical observation that the rotation rate of the Earth as dynamically measured was the same as that kinematically observed relative to the "fixed stars," the fact that, according to Mach, strongly suggested that the irrotational frame was determined as the fixed-star frame. In the new version the Machian once again asks us to consider the non-rotation of the cosmos not as a contingent fact but as defining what absolute non-rotation really is.

The claim of Barbour and Bertotti to explain the existence of inertial frames in a deeper sense becomes much more plausible when their new variational method is introduced. Applying this variational principle to the cosmic system demands that this system have no rotation as a consequence of the basic dynamical law. From that point on the existence of the usual inertial frames relative to which the motions of isolated systems are described by the usual Newtonian dynamics is assured from the cosmic lawlike non-rotating frame, just as the Newtonian absolute time for these individual systems follows from the cosmic ephemeris time. So here is, surely, a deeper account of absolute motion than the mere positing of inertial frames in the usual Newtonian interpretations.

But is the price paid for this deeper explanation too high? Should we accept this alternative account as being preferable to the old Newtonian account or any of its variants? Much remains open here.

The Barbour–Bertotti account is of necessity an account that takes the scientific characterization of the cosmic whole as prior to understanding the behavior of the local, isolated systems that are the typical subjects of theoretical explanation in ordinary Newtonian dynamics and, indeed, of most other fundamental theories. In this new approach cosmology must come first. For the fundamental variational principle implies no rotation for the system it describes, and few fragmentary, isolated systems meet this condition. Perhaps the cosmos does. And, insofar as it does, we have a new version of the argument offered by Mach to the effect that this new account explains why the cosmos must have no rotation, whereas that fact remains unexplained in the older Newtonian theory.

One aspect of this "cosmic" theory deserves comparison with a different invocation of a "cosmic" solution to a problem in fundamental physics. In the kinetic-theory/statistical-mechanical attempts to ground thermodynamics on the micro-constitution of objects and the dynamics of these micro-constituents, one deep problem has always been the source of the time asymmetry of the Second Law of thermodynamics. How can such a time-asymmetric principle as the increase of entropy of isolated systems, or at least the statistical non-decrease of entropy of such systems, rest on an underlying theory with time-symmetric dynamical laws? Another problem is the very existence of a world in which non-equilibrium states are pervasive. This is a problem since the statistical-mechanical theory postulates that equilibrium is the overwhelmingly most probable state for any system. Faced with this second problem, Boltzmann himself offered a "cosmological" way out. He suggested that the cosmos was, overall, in equilibrium, but that we happened to live in a temporally and spatially restricted part of the cosmos that had fluctuated away from equilibrium. He also offered an argument that would later be called "anthropic." This was to the effect that only in such a non-equilibrium fragment of the world could we exist as observers.

One contemporary solution to these problems is to trace the ultimate thermodynamic asymmetry and the pervasiveness of non-equilibrium back to the special nature of the cosmos at, or just after, the initial singularity of the cosmos, the Big Bang. Assume matter is thermalized in this state, that is, that it has high entropy. But assume that space itself (or the spatial distribution of the matter) is uniform at this time. Given the purely attractive notion of gravity, this can be considered an extremely low-entropy condition. From then on we have gravitational clumping of matter leading to a non-uniform spatial distribution of matter. This leads to the "hot stars in cold space" of our familiar Universe, a non-equilibrium condition from which temporarily isolated systems can be extracted in low-entropy states, systems which then proceed to higher entropy in the future time direction. There are many problems with this explanatory account of the thermal time asymmetry of the world, but we cannot pursue those here. In particular, whereas this account would help greatly with explaining the cosmic pervasiveness of non-equilibrium, more is needed to get the desired time asymmetry of thermodynamic processes of isolated systems in the world.

What should gain our attention, though, is a sort of parallel issue to that faced by Barbour–Bertotti cosmic Machian dynamics. The cosmic solution to the thermodynamic problem posits an initial state for the Universe as a whole that is, in statistical mechanics' own terms, highly *improbable*, since it is of enormously low entropy and according to the very theory itself it is high-entropy states that are overwhelmingly probable. Compare this with the situation in Machian dynamics where we are asked to posit a state for the cosmos as a whole, having absolutely no spin whatever, something that is rarely, if ever, encountered among the usual, temporarily isolated, dynamical systems that are fragments of the cosmos.

Why do we have such a "low-probability" initial state for the cosmos in the thermodynamic case? Here speculation runs rampant. Typical are thoughts about there being an infinite number of universes in all sorts of entropic conditions, or ours being improbable, but with the necessity of such low entropy for the Universe to have observers within it. In other words, we are confronted with one version of the notorious (and misnamed) anthropic principle. But another suggestion (due to R. Penrose) is that there is a lawlike constraint of "white holes" imposing that they have such low entropy. Since the Big Bang is the only white hole we know of, this "law" is, of course, hard to test. Compare this with the neo-Machian's claim to the effect that the "improbable" non-rotation of the cosmos is the consequence of the fundamental cosmic variational law that grounds all of dynamics.

The deep epistemological problem in both stories is, of course, the fact that, as Mach said, the Universe is given to us once only. Or, to use C. S. Peirce's phrase, universes are not as common as blackberries! If the cosmic rotation really is null, then, by design, the traditional Newtonian account or its neo-Newtonian revisions will have all the same empirical consequences as does the Barbour–Bertotti neo-Machian account. This leaves us with the issue as to how we could possibly make a rational decision to accept the one account or the other.

Some versions of positivism might argue that the two accounts are really not two distinct theories of the world at all. Mach sometimes talked that way about his "fixed-stars" version of absolute space versus that of Newton. Some *possible* observation could distinguish the theories – finding the cosmos in absolute rotation for example. But if the cosmic matter isn't rotating, such a "possible" observation remains just that. And for a radical positivist such as Mach sometimes takes himself to be, equivalent empirical prediction in this world as it is is enough for full theoretical equivalence.

Otherwise we may hope that some future development of foundational dynamics will give us deeper insight into which account ought to be taken as correct. Invoking general relativity won't help, since there are the standard versions of that theory that can be shown to have distinctly non-Machian aspects, but Barbour–Bertotti have constructed neo-Machian versions of general relativity as well. The epistemological issues remain as open as the question of which of the alternative accounts we ought to think of as the true foundation of dynamics.

SUGGESTED READING

The Barbour–Bertotti reformulation of Machianism can be found in Barbour and Bertotti (1977) and in Barbour and Bertotti (1982). A fuller development with many additions and reflections is as yet unpublished (Barbour, *Absolute or Relative Motion*, vol. 2, ms.). A rich collection of articles outlining Machian perspectives old and new is given in Barbour and Pfister (1995). For a technical discussion of the effects of rotation on the remnant radiation see Su and Chu (2009) and the references cited therein. The "cheap" way out of substantivalism is discussed in Sklar (1974), pp. 229–234. Outlines of the role played by the cosmic initial-low-entropy postulate in grounding explanation in statistical physics can be found in Chapter 8 of Sklar (1993), Chapter 4 of Albert (2000) and Chapter 4 of Price (1996).

CHAPTER 21

Modes of explanation

21.1 THE VARIETY OF EXPLANATORY MODES IN DYNAMICS

We have seen that dynamics is a theory with a multiplicity of explanatory structures that can be used to formulate its lawlike conclusions and to provide explanations of the behavior of systems within its purview. We have also seen that the threads of some of these structures can be traced back to the earliest days of the development of the theory.

One pattern of explanation in dynamics we might call the "Newtonian." Here one must first posit an appropriate structure that admits a preferred metric of time and the existence of the preferred inertial reference frames to which all motion is to be referred. Inertial motion with constant speed and direction is taken as the "natural," "unforced," state of a body.

Inertial mass and force are introduced. The former is an intrinsic property of a piece of matter representing its resistance to having its state of motion changed, and the latter is the (vector) measure of the influences that can change the state of motion of a system. The fundamental law, of course, is the proportionality of the linear momentum change of the system to the force applied to it. This initial Newtonian framework must be supplemented, as was first seen by Euler, by a corresponding notion of moment of inertia as intrinsic resistance to change of state of rotation and of moments of forces (or torques) as the measure of the influences generating changes of angular momentum.

Fundamental to the theory are the principles of "reciprocity." The first of these was realized by Newton in his Third Law of action and reaction, and later supplemented by a corresponding principle for changes in angular motions. The corresponding principles of conservation of momentum (knowledge of which, of course, antedated Newton) and conservation of angular momentum were later seen to be reflections of the basic symmetries of the spacetime presupposed under translation and rotation.

The second pattern originated with James Bernoulli's work on the compound pendulum. Here the resources developed in statics from ancient Greece onward are deployed for dynamics. The principle of balancing not just forces at a point but moments of forces as well is crucial in statics, as is the observation that the ratios of the lever-arm of the moments are proportional to the ratios of the "virtual motions" of points when the system is slightly displaced from static equilibrium. Bernoulli's great contribution is to generalize this to dynamics by adding the negative force of mass times acceleration to the static forces for systems in motion.

This work was generalized by d'Alembert in his presentation that tries to eliminate the notion of force altogether in favor of "motions," and brought to full generality and its current form by Lagrange. It is Lagrange also who realized the full power of the method of "virtual work" or of "d'Alembert's principle." Move to a description of the system in terms of generalized coordinates and velocities. When the constraints on a system are holonomic, when, that is, they are presented as functionals of the original Cartesian coordinates, a great simplification of the description of the system is possible. Even more simplification follows when all the forces are conservative. One can drop all reference to coordinates where no virtual work is done in the motion, and one can drop all explicit reference to forces altogether. The result is the famous Lagrange equation framed in terms of the kinetic and potential energies alone.

Finally there is the explanatory approach that, taking off from the work of Hero and Fermat on geometrical optics as derivable from a least-time principle, in the hands of Maupertuis took on the first version of an extremal, least-action, principle as fundamental for dynamics. Maupertuis' work was made mathematically respectable by Euler and later by Lagrange. Later still one gets revised and independently interesting new versions at the hands of Hamilton and Jacobi.

In all of the refined versions of the theory of dynamically possible motions as motions that extremize action along the dynamical trajectory there are invariably restraints placed on the kinds of dynamical systems to be treated by the method. If one is dealing with constrained motion, the constraints are taken to be holonomic. And, in all the standard textbook accounts, the forces are, once again, taken to be conservative, that is to say derivable as derivatives of a potential that is at most a function of the generalized coordinates and velocities. And, of course, the net result of the use of the extremal principle as fundamental is cashed in in terms of the derivation of Lagrange's equation as the differential, necessary and sufficient condition, for the path integral to have its extreme value.

In the nineteenth century this work developed into the grand explosion of what is sometimes called "analytical dynamics," that version of dynamics in which force and torque play no explicit role but where the starring place is reserved for energy. This is, of course, hardly a return to Leibniz, since the place of the potential-energy function in this work goes far beyond that of merely making conservation of energy a principle of the theory.

Perform a Legendre transformation on the Lagrangian of Lagrange's theory and one gets the Hamiltonian which, for conservative systems, is just the total energy. Define generalized momenta conjugate to generalized coordinates, once more using the Lagrangian in the definition. Then one can write the fundamental dynamical laws in terms of sets of pairs of first-order differential equations instead of sets of single second-order equations, the Hamiltonain equations. Go a little further and one can get the Hamilton–Jacobi partial differential equation for the action, allowing one to characterize whole classes of trajectories from distinct initial states in one blow. With ingenuity one can make the treatment of dynamics even more symmetric between generalized position and momentum by using Poisson brackets as the form in which to frame the fundamental equations of motion.

In the late nineteenth and twentieth centuries analytical dynamics became richer still with the application of the power of mathematical analogy and abstraction. Abstract spaces such as configuration space, contact space (configuration space supplemented by time) and phase space were introduced. Put a natural metric on configuration space and trajectories in it become geodesics of a non-Euclidean geometry. Note that the conservation rules impose a conserved phase volume on phase space as dynamics shifts its points about and a new kind of "geometrization" of dynamics, as mappings of points in a symplectic space, becomes available. Focus on the algebraic relations among the Poisson brackets and the abstract apparatus of Lie algebras can now be applied to dynamical problems.

The payoff of all this work is multiple. For the working physicist one obvious payoff is that with this enormous repertoire of methods in dynamics one is able to solve a vast array of specific problems. Even problems of simple systems – small numbers of interacting point particles or asymmetric rigid bodies, say – cannot be dealt with as a practical matter by the methods of forces and torques. Once constraints come into play and once one deals with such complex matters as interacting fields, the methods of analytical dynamics become indispensable for actually solving problems.

The new concepts and methods prove indispensable also when dynamics functions as a component of other theories. For example, the picture of dynamical change as a volume-conserving shift of the points of phase space provides the central device for dealing with systems in equilibrium in the version of statistical mechanics that begins with the work of Maxwell and Boltzmann on what the latter called "*Ergoden*," and continues on to Gibbs' famous work on "ensembles." Further, the variety of methods of analytical dynamics plays an ongoing and crucial role in the formulating of new theories to replace the Newtonian dynamics with which we are concerned. Hamiltonian dynamics and Poisson brackets are the key to basic quantum theory. Lagrangian methods are the basic framework in which quantum field theories are expressed and extremal methods are the key, in the hands of Dirac and Feynman, to generating the perturbation theory needed to solve concrete problems in that theory.

Within Newtonian dynamics, as we saw in Chapter 18, the panoply of methods available within analytical dynamics provides one insight after another into structural features of a world described by Newtonian dynamics. These features might never have come to light were it not for the multitude of transformed modes of description and explanation provided by the newly developed mathematical accounts of the theory.

There are, however, certain questions we might ask about the place of these methods in our overall explanatory account of the world. Some of these arise out of issues "internal" to the structure of dynamics itself. Others arise out of general debates about the nature of explanation that exercise philosophical methodologists.

The primary questions concern the legitimacy of the analytic methods to lay claim to a possible role as the "fundamental" explanatory structure for the theory. Could we justifiably take the extremal principles, for example, as the ground explanations of the theory? Or is it the case, rather, that these extremal methods, and all the other methods of the analytical approach, must be viewed in the light of "derivative" principles, relying for their explanatory legitimacy on the fact that they can be inferred from the "true foundations" of the theory, foundations in the Newtonian notions of force and its effect on changing states of motion of otherwise inertial systems?

We will first look at the way in which these questions take on a particular form when the issues are those that arise primarily out of the special structure of dynamics itself. Then we will look at the broader philosophical concerns.

21.2 INTERNAL ISSUES CONCERNING ANALYTICAL DYNAMICS

Could one legitimately take one of the extremal principles, a least-action principle, as the fundamental postulate of Newtonian dynamics? Could one take one of the other standard formulations of analytical dynamics as fundamental, say Lagrange's equation, Hamilton's equations, Poisson brackets or Hamilton–Jacobi theory?

One fact immediately stands in the way of such a proposal. All of these principles are presented as applicable, in their standard forms, only under restrictive conditions regarding both the nature of constraints on the system and the nature of the forces involved in the system's dynamics.

For constrained motions, it is easy to put the constraints into the least-action principle (by adding suitable functions multiplied by Lagrange multipliers to the usual action integrand) in the case where the constraints are holonomic. We saw one application of this in the last chapter, where Barbour–Bertotti Machianism can be formulated in terms of an extremal principle with zero angular momentum about the center of mass of the Universe and zero energy imposed as constraints within the principle. But when constraints are not holonomic, it is impossible to include them in a standard extremal principle.

Should someone who wants to take analytical dynamics as fundamental worry about this? The easy reply is obvious. Constraints are not usually like those used in the Barbour–Bertotti theory that fix a constant of motion to a particular value. Rather, they are devices that work by idealizing some real system of internal forces in the system. By making this idealization one can often get away with ignoring some degrees of freedom of the system and vastly simplifying the problem solving involved. But, at bottom, constraints, certainly non-holonomic constraints, are not part of "fundamental reality."

A rigid body, for example, is not really perfectly rigid. By pretending that it is we can ignore the internal forces binding the parts of the object together and the very slight motions of these parts relative to one another when we do the dynamics of the whole object taken as rigid. The same goes for a "bead on a wire" type of constrained dynamics. And for such non-holonomic cases as the famous disk rolling without slipping on a plane.

So, if we eliminate the notion of the usual constraints as mere idealizations of convenience, we could still argue that analytical dynamics can be taken as the foundational posit for dynamics, a dynamics that deals "at the base level" only with unconstrained systems and their full panoply of internal forces.

But there is a much deeper problem with taking analytical principles as fundamental. In the usual textbooks, the analytic methods are applied only to systems whose forces are conservative, that is, systems where the forces can be given as gradients of potentials. The potentials can be functions of generalized position and of generalized velocity, but forces not derivable from a potential at all are left out of consideration. It is even stated sometimes that the extremal principles and their consequences can be applied only where the forces are so derivable from potentials.

The question of the limits of the applicability of analytic methods was raised in the nineteenth century by Jacobi. An elegant result of Helmholtz provided an important answer to the question raised. One takes a dynamical path followed by a system and varies it. Relative to the variation and set of forces a variational form can be defined that depends on the functional form of the forces acting in and on the system. A system of variational forms can be such that any two variations in it can bear a relation to one another of "adjointness," if the difference between the product of the first variation times the second's form and the second variation times the first's form is the derivative of a two-place function of the variations. If the forces are such that the two variations of the system with the same forces acting are adjoint to one another, the system is called "self-adjoint." Helmholtz was able to show that the condition of self-adjointness is what is needed in order for the dynamics of the system to be represented in the standard modes of analytical dynamics.

If one moves from the presentation of the results in terms of variations of integrals, the mode of representation needed for extremal principles, to the mode given by Lagrange's equations as solutions to the extremal problem, then elegant simple conditions on the forces involved can be derived that tell one whether or not the system can be represented in the usual analytical format. If one restricts one's attention to systems with at most holonomic constraints, and if the forces are all local, given as functions of the variables rather than as integrals over them, and if the forces are sufficiently smooth (non-impulsive), and if a technical condition of "non-degeneracy" is met, then self-adjointness becomes the condition for representing the dynamics of the system in analytical form (say by means of Lagrange's equation with no residual force term on the right-hand side).

Forces that are non-conservative can satisfy the condition of self-adjointness. These can include forces that depend explicitly upon time, forces that are non-linear in generalized position, forces that are non-linear in the generalized velocities and even forces that depend linearly on the accelerations of the components of the system.

But what about forces that don't meet the preconditions (smoothness and locality, for example) or that are of a form such that they could not be self-adjoint (non-linear in the accelerations, for example), or that are just non-self-adjoint? Doesn't all of this show that the range of application of analytical dynamics is a proper subset of the range of dynamics in general? And doesn't that show that analytical principles cannot be foundational?

But it isn't that simple. For one thing, one can generalize the scope of analytical dynamics. Ruggero Santilli, for example, proposes the "Birkhoffian generalization of Hamiltonian mechanics." This includes a generalization of the usual extremal principles that replaces the old "momentum times generalized velocity" term in the integrand in Hamilton's version of the extremal principle by a new term involving a function of the generalized coordinate multiplied by the generalized velocity. And it complicates the Hamilton equations by adding terms to the usual left-hand side of the equation.

This generalization can be shown to have remarkable features. With this apparatus in place it can be shown that any finite-dimensional system that is free or holonomically constrained, whose forces are functions only of time, position and velocity, and whose force functions are local, smooth and non-degenerate, can be given an analytical representation in terms of the new extremal principle and the new Birkhoffian equations. This is true even if the force system is not self-adjoint.

What is more, the new apparatus carries with it some of the nice geometric and algebraic structures that made the older analytic dynamics so flexible and elegant. A two-form that is like the symplectic form of ordinary analytical dynamics exists, and brackets can be defined that form a generalized Lie algebra just as the Poisson brackets of the older theory form a Lie algebra. But phase space must be replaced by "dynamical space," and some of the familiar results of the older theory no longer go through. This is hardly surprising. A price must be paid for going beyond what were the provable limitations on the representational capacity of the older analytical dynamics.

Can one go further still? Only by giving up even more. One can look for more general structures that might encompass even non-local interactions, but where the force function is still a function only of time, position and velocity and not of acceleration or higher derivatives of the motion. "Symplectic admissible" geometries can be constructed as generalizations of symplectic geometry and "Lie admissible" algebras as generalizations of the Lie algebras of the older brackets. The underlying dynamical space, though, becomes a cloudy issue when non-locality comes into play. And many of the nicest aspects of the standard structures must be dispensed with.

From the point of view of someone trying to solve actual problems in dynamics, the point of this exploration of what is called the "inverse problem" in dynamics is clear. The methods of analytical dynamics, where they can be applied, provide a rich range of resources for solving particular problems that would be intractable using the standard Newtonian formalism. Even if a problem is outside the range of those that can be treated by the standard methods of analytical dynamics, perhaps these now discovered extended methods (Birkhoff equations, dynamic spaces, generalized symplectic forms and generalized Poisson brackets, symplectic admissible forms and Lie admissible brackets) can be applied for equally useful purposes of problem solving.

But there is methodological interest in this program and in its results as well. What does all this have to tell someone who is interested in the following questions. What ought we to take as the fundamental dynamical posits of dynamics? Could an extremal principle serve as a fundamental posit, or must it be considered merely a clever derivative device whose place can never be at the base of the theory?

At first glance the results seem to tell us that only the Newtonian formulation can be fundamental, not the devices of analytical dynamics. After all, the whole program relies on taking forces as basic and then asking under what restrictive conditions the force equations can be transformed into the modes of analytical dynamics. Doesn't that clearly indicate that force (and maybe torque) are primitive in any fully general version of the theory? Even more, the results of the study of the inverse problem clearly show that there are severe limits on what standard, or even generalized, analytical dynamics can do. If forces are non-linear in the accelerations, if they are impulsive and so fail to meet the conditions of analyticity imposed on the theorems about analytical dynamical representability, or if they are non-local, then analytical dynamics cannot deal with them.

Doesn't this show that even from an internal perspective, looking only at dynamics as a theory and ignoring philosophical issues about what count as legitimate basic explanations, analytical dynamics, extremal principles say, can't count as fundamental to the theory.

Once again, it isn't that simple. In an introductory chapter to his study of the inverse problem Santilli poses the speculative possibility that forces not treatable in the standard way by ordinary analytical dynamics might play a role in foundational physics. They certainly play a role in the kind of physics familiar from continuum mechanics where impulsive forces (shock waves), contact forces that can't be derived from potentials and might be non-self-adjoint, and even non-local forces may be introduced to deal with the special constitutive features of matter.

But there is a long-standing debate about just how seriously such forces should be considered as a real part of the world. There is a piece by Duhem in his role of anti-atomist where he praises the ability of macroscopic continuum mechanics to deal with such things as capillary attraction, and mocks the attempts of those who try to give a reductive account of it in terms of atoms and purely potential derived forces. Those who deal with macroscopic matter in all its messy forms will be skeptical indeed that all can be reduced to the cases where the methods of analytical dynamics reign sufficient.

But when one enters the realm of foundational physics, forces seem to have vanished as explicit elements of the explanatory scheme. The current foundational theories are not, of course, framed in Newtonian dynamics. They must be relativistic and they must be quantum mechanical. No matter. These successor theories carry on enough of the framework of the Newtonian theory for our purposes.

When we look at field theories, we find Lagrangians expressing the dynamics of fields and their interactions. When we look at quantum mechanics we find "canonical" equations utilizing the Hamiltonian. When we look at quantum field theory we find the extremal methods of Dirac and Feynman at the basis of perturbation theory. When we look at statistical mechanics we find phase space with its volume conserved under dynamical transformation given by a symplectic form. It is the apparatus of analytical dynamics and not that of the Newtonian version of the theory that is the conceptual framework for these theories. Implicit in this is the assumption that the kinds of interactions that would lead to a breakdown of the analytic methods just don't occur at the level of fundamental physics (despite Santilli's suggestion that they might someday so appear).

Here is one way of looking at this. It is, to be sure, very useful for understanding the way dynamics functions as an explanatory theory to make a distinction between the fundamental dynamical laws and the constitutive equations that delineate the particular interactions systems have with one another. One must find the constitutive features – the particular forces, the particular potentials, or the particular Lagrangians or Hamiltonians – and "plug" them into the general dynamical posits in order to set up the problem for solution.

But our very idea of what those fundamental dynamical posits ought to be may be dependent upon our beliefs about what the range of constitutive behaviors is like in the world. If we believe that "peculiar" forces are endemic at the fundamental level, we will opt for the Newtonian scheme as foundational for the dynamics. But if we believe that in the world at the level of foundations there are no interactions but those that can be dealt with by an extremal principle or by its consequent structures of Lagrangians, Hamiltonians, phase space, symplectic structures and Lie algebras of Poisson brackets, we might then very well opt for taking extremal principles as "the" fundamental dynamical laws.

From the latter perspective we might come to think of the use of "force" as merely an intermediate step. It was a useful concept for understanding dynamics when our understanding was limited to the behavior of macroscopic objects with all their "messy" frictional, impulsive and non-local characteristics. From that older understanding of dynamics the methods of analytical dynamics might be derived as principles holding for a limited range of possible cases of dynamic systems. But in the end our deeper understanding of the constitution of the world might lead us to maintain that it is the analytical methods, the extremal principles in particular, that are fundamental and that capture the "essence" of the dynamics of the world.

Here it is useful to remember that Maupertuis' first crude introduction of a least-action principle predated Euler's making it mathematically respectable and Lagrange's generalization of Euler beyond the case of the single particle. And it long predated Lagrange's expressed relief that something so philosophically notorious as an extremal principle could be derived from the more "respectable" Newtonian formalism or the formalism of virtual work generalized to the dynamical case.

But none of this deals with the lingering philosophical worry that there is something wrong "in principle" about taking an extremal principle as foundational at the level of explanation. It is to those worries that we now turn.

21.3 EXTREMAL PRINCIPLES AND THE PHILOSOPHY OF EXPLANATION

The philosophical objection to allowing extremal principles the role of fundamental explanations is intuitive and this objection has been repeated throughout history. These explanations explain processes over a duration of time, unlike the "causal" explanations that explain what happens at a single time. And, worse yet, the explanations they offer of what happens over some time interval require reference to what is going to happen at some time future to that interval. For we derive the least-action path, say, by reference not only to the initial state of the system but to its final state as well.

Our common-sense everyday explanations are, for the most part, causal. Why did an event occur? Because something happened in the past that forced, made, caused that event to occur. But many of the explanations in science don't seem to be causal. There are functional explanations in biology. Why does the heart beat? In order to pump the blood. And in physics we have explanations by laws of coexistence – the ideal-gas law ($PV = k_B T$) for example. And we have the extremal principles in optics and in dynamics.

The nature of scientific explanations has been one of the persistent problem areas in methodological philosophy of science. Much of the debate has centered around the following issue: Is it sufficient for explaining an event to show that that event is connected to other events by a law of nature? Does subsuming what happens under a lawlike regularity of happenings suffice for explanation? The formalization of the idea that to explain is merely to subsume under lawlike correlations is usually called the "deductive–nomological" model of explanation. It comes in strictly lawlike and statistical, probabilistic, versions.

What inspires this notion of explanation? Part of the inspiration comes from a reading of Hume as telling us that there is nothing to causation except spatio-temporal contiguity, constant conjunction and a psychological habituation on our part. The idea, then, is that the remnant of causation left in the world is just subsumption of events under generalities connecting kinds of events to other kinds of events. But part of the inspiration for the deductive–nomological views, most certainly, was the appearance in physical science of laws governing the regularity of physical behavior that did not seem causal in nature. Extremal principles were one such type of non-causal law.

There are many critics of the deductive–nomological model. Most criticisms start from examples of subsumptions under a law that don't seem explanatory. We can explain the length of the shadow of a building using the building's height and the position of the Sun, but we can't explain the height of the building from the length of the shadow and position of the Sun, even though the lawlike connection works both ways. One very common proposal is that lawlike subsumption in order to be explanatory requires an additional causal element.

But is that so? And if it is at least sometimes so, what does it mean to add a causal element to the explanatory picture? For our purposes it is most important to focus on that aspect of the notion of causation that takes causation to be deeply asymmetric in time: Causes precede their effects.

This was a condition Hume demanded for causation along with his conditions of proximity, constant conjunction and the psychological element of expecting the effect when encountering the cause. But why do we think of causation in this time-asymmetric way, with the past determining the present and the present the future – and never the other way around?

One general class of answers seeks the time asymmetry of causation in some deep "metaphysical" aspect of the nature of time itself. We think of the past as both having determinate reality (events definitely occurred or they did not) and of being immune from any kind of change. And we think of the future sometimes as being open to change, depending on what is done now, and sometimes, in a manner going back to Aristotle, as not having any determinate reality at all. The latter means, roughly, that a statement as to what will happen in the future does not now have any truth value at all.

Sometimes the "metaphysicians" talk about "branching time worlds." Think of the world as a tree with a single trunk into the past but many branches of possible later universes going into the future. As time progresses the single past grows, always having at any time a manifold of possible futures as branches coming after what is then the "now."

Causation, then, is asymmetric in time because of this deep metaphysical asymmetry in the nature of being itself in time. Of course, it must be the past that causes the future and not the other way around. The past is a realm of determinate happenings that could very well force the future to come into being in one way or another. But the future has no determinate reality at all, no genuine being in fact. So how could it possibly be determinative of the present or past? Unsurprisingly, views of this kind will lead one to think of the extremal principles as derivative rather than foundational in dynamics. The Newtonian mode of forces determining changes of motion fits easily into a "causal" picture of explanation. This is so even though on reflection one can see that a force-type account might very well admit a time-symmetric, lawlike subsumption of phenomena. But the extremal mode of looking at things sits very uncomfortably as a fundamental pattern of explanation from this perspective of a deeply time-asymmetric notion of causation embedded in a time-asymmetric picture of being itself.

But there is quite a different way of looking at the origins of our ideas of the asymmetry of time and the asymmetry in time of causation. What can we find, not in metaphysics, but in fundamental features of the world as described by physical science itself that might inform us of the origin of these ideas of temporal and causal asymmetry? One general program that seeks to answer this question takes a complex and subtle route from the theory of heat to an alleged physical answer to the question of the origins of our intuitive time asymmetries.

Thermodynamics, the theory of heat, was known from the eighteenth century on to contain a fundamentally time-asymmetric component. Heat flows from a hot body to a cooler body, not the other way around. Other time asymmetries exist. Bodies of fluids that are asymmetric spatially in density flow to even-density bodies if left alone, not the other way around. All of these asymmetries were captured by the notion of entropy, a measure of the uniformity of state of a system, and a claimed "Second Law of Thermodynamics" to the effect that in isolated systems entropy increases into the future. Other related physical asymmetries were later noted. For example, accelerated charged particles emit coherent radiation that spreads into space. But we don't see coherent radiation coming in from space to converge on a charged particle and accelerate it. A more homely example is that of ripples in a pond when a stone is thrown in. We don't see converging ripples from the shore picking up stones and throwing them out of the pond.

But why do systems show these pervasive asymmetries in time? Deep puzzles remain about this, but there is a large degree of agreement on part of the story. Look at macroscopic systems composed of a vast number of microscopic components, say a gas composed of molecules. The micro-components move about according to the laws of dynamics, say the very laws of Newtonian dynamics we have been exploring. Somehow we must then account for the macroscopic entropic asymmetry of the macro-system in terms of this dynamic microscopic behavior. But how can we do this if the laws governing that microscopic behavior – say the Newtonian laws – are demonstrably time-symmetric in their account of motion and its causes, allowing for any behavior that is possible the exact time-reverse of that behavior as possible?

The most widely accepted, but tentative, answer to this question is to seek the explanation of the entropic asymmetries in a fundamental asymmetry between the initial and final states of temporarily isolated systems. And a widely accepted, but even more tentative, explanation for that traces matters back to a claimed profound fact about the state of our cosmos at (or just after) its inital cosmic singularity – the Big Bang. The claim is that if we assume a very improbable (in statistical-mechanical terms) initial state of spatial uniformity for the cosmos at the Big Bang – and an initial very-low-entropy state given the special nature of gravity as a force – we can then trace the initial-low entropy states of systems temporarily isolated from the cosmic system back to this initial super-low-entropy cosmic state. And we can then use these initial-low entropy states to explain the parallel-in-time increase of entropy of the so-called "branch systems." And there is evidence from the remnant black-body background radiation of the cosmos that such an initial cosmic smoothness really was the case.

A truly satisfactory account of the origin of the entropic and entropic-like asymmetries in time of physical systems, an account which eschews any metaphysically inspired fundamental asymmetry of time itself, remains elusive. But let us take it as given that such pervasive entropic time asymmetries of physical systems exist. How does that explain the existence of our strong intuitive feelings that time itself is deeply asymmetric and that causation itself is asymmetric in time? Here we can only, once again, cursorily glance at some of the proposals that have been made in order to try to answer that question.

What about our intuition that the past is determinate and fixed and the future indeterminate and open to many possibilities? One way of trying to answer this question from the entropic standpoint is to seek the source of the intuitions in the asymmetry in time of our records, including our memories. We have a kind of epistemic access to the past that we don't have to the future. This is not to deny that we can sometimes know what the future will be like, and sometimes cannot know what the past was like. Rather it is to emphasize a distinct mode of evidence about the past – having memories of the past or having records of what went on. Then one goes about trying to explain how the asymmetry of records can be accounted for in terms of the entropic asymmetry. This part of the account, however, has proven a very intractable problem.

The program of trying to find an entropic account of memories and records, and hence an entropic account of the "fixity" of the past, is entangled with the program of seeking an entropic account of the intuitive asymmetry in time of causation. Why do we think of causes as preceding in time their effects? Some attempts at dealing with this issue have taken as their starting point the way in which, in many familiar situations, changes in some "compact" feature of the world are correlated with changes in some "spread-out" feature of the world in a time-asymmetric way. Once again, think of the compact event of the pebble dropping in the water correlated with the later, spread-out, coherent wave of ripples in the pond; or the compact acceleration of the charged particle associated with the later, spread-out, coherent radiation coming from the charge.

Hans Reichenbach lucidly introduced the notion of an earlier event "screening off" later correlations of events. Let A be the earlier event and B and C the later events. Suppose the chance of B and C together is higher than one would expect given the chances of B and C independently. We often find that if we look instead at the conditional probability of B given A and C given A we once more find independent probabilities. The earlier event A, which is intuitively the common cause of B and C, "screens off" the excess correlation found between B and C. Yet it is unusual (although not unheard of) to have an event to the future of a pair of correlated events screening off their correlation. The suggestion would then be that we think of "causes" as screening-off events, and that the time asymmetry of causation arises from the time asymmetry of screening off. And the hope is that this time asymmetry finds its home in the entropic asymmetries of physics.

Another variant comes from David Lewis. Causes are thought of as having their roots in counterfactuals. A is the cause of B if (among other things), if A had not occurred, B would not have either. And we think of counterfactuals in a time-directed way. "If the past had been otherwise, the future would have changed" seems right to us, but not the other way around. But why do we read counterfactuals in this time-oriented way? Could it be that when we think of "what would have happened" we imagine localized changes in the world, compact "miracles." Then, imagining the pebble not hitting the water, we think the ripples would not have spread. But the "miracle" needed in the other time direction would be the spread-out miracle of all the ripple effects not being the way they are in the present and then tracing that by the laws of nature back to the past. Once again, could entropic asymmetry in time lie at the heart of causal asymmetry in time?

One frequent objection to theses of this sort is that there are lots of cases where nothing like entropic spreading is going on, but where we still think of the causes as preceding in time the effects. But there is a reply to this. It works by arguing that our basic ideas of causation and its temporal asymmetry come from the cases where entropic considerations are playing their role. We then "project" the basic asymmetry in time we see there, which we then think of as the "causal" asymmetry, to cases where the interaction which occurs is governed by the laws of nature but where entropic aspects are completely lacking.

This is just the barest outline of an extensive and certainly controversial program. Focus on the entropic asymmetries of systems in time and on related time-asymmetric processes such as the radiation asymmetry. Seek the origin of such asymmetries in the fundamental laws of processes supplemented as needed by probabilistic assumptions governing the initial conditions of systems. Next, perhaps, seek out the origin of the time asymmetry of systems in a fundamental "cosmic" asymmetry of the "initial condition" of the Universe. Add to this internal physics program a "philosophical" program that proposes to find the ground of our intuitive ideas – that time itself has a fundamental asymmetry with the past fixed and the future open and that causation is time-asymmetric with causes always antecedent to their effects – in the time-asymmetric entropic processes of physics. Finally use these results to disabuse us of the idea that the intuitive asymmetries of time and causation are based on some profound "metaphysical" nature of time or causation that lies too deep for the physical theory to explain.

What concerns us here is what the consequences would be for our understanding of the explanatory place of the extremal principles if this account of the origin of the intuitive time and causal asymmetries is correct. From the very introduction of extremal principles into physical explanations (least action in dynamics, least time in optics) the concern has been expressed that such explanations cannot possibly be taken as legitimate; or, at least, that their legitimacy can rest only on the fact that they can be derived from some causally respectable theoretical underpinning (force and acceleration in dynamics, wave propagation in optics). But if the "entropic" account of the intuitive time asymmetries is correct, the possibility of taking the extremal principles as explanatory at the fundamental level seems less problematic.

From this perspective the special "metaphysical" aspects of time and causation are something of an illusion. Those intuitive ideas of some deep asymmetric nature to time and cause that goes beyond specific features of the world in time and as governed by lawlike regularities, probabilistic distributions over initial conditions and a special nature of the cosmic initial condition reflect only the pervasiveness of entropic features in such deep aspects of our lives as the very existence of memories and records of the past and the special natures of the kinds of interventions we can and cannot make in the manipulation of systems. But if that is all there is to the asymmetries, there seems little reason to deny that the extremal principles, despite their violation of the intuitions of their non-causal nature and their explanation of processes in a global way that adverts equally to past, present and future, could be taken as fundamental explanatory principles of our theory. From this point of view it seems unnecessary to give them credentials of legitimacy by deriving them from some "causal" principles that are more acceptable at the foundational level.

Here it would be worthwhile to emphasize again the fact that fundamental physics seems to be full of explanations that cannot be forced into the "causal" form: Explanations of coexistence, the extremal principles and laws of "entanglement" in quantum mechanics are some examples. As we have noted, the appearance of such principles within fundamental science was one of the crucial reasons why methodologists of science were encouraged to drop the notion of causation entirely when looking for some general model of scientific explanation and replacing it with subsumption under general laws, whether or not these laws could be given a "causal" reading.

In order to answer a question internal to dynamics, the question of the legitimacy of extremal principles as fundamental explanatory modes, we see that we must range beyond dynamics itself. Dynamics plays a crucial role in the theory of statistical mechanics that is taken to underlie the principles of thermodynamics and its entropic claims. But dynamics isn't all there is to that statistical theory. The degree to which the fundamental probabilistic posits of the theory can be extracted from the dynamics of the micro-constituents of systems is quite controversial. But the need for some additional posits about initial conditions seems necessary. We cannot really decide on the legitimacy of the extremal principles as the foundational principles of dynamics without understanding where our ideas of causation, the time asymmetry of causation and the role of causation in explanation come from. And to do this we must examine not just dynamics but the totality of our best available foundational science. If those who find the origin of the intuitive asymmetries of time and causation in the asymmetric, entropic behavior of physical systems in time are right, and if they are right that these asymmetries can be accounted for without resort to alleged metaphysical asymmetries of time itself or of some primitive notion of time-asymmetric causation, then there seems little to hinder letting a dynamics founded on extremal principles as fundamental be counted as just as methodologically legitimate as a dynamics founded upon a causal-sounding notion of force engendering acceleration. All of this, of course, subject to the issues of the generality of extremal principles discussed in the earlier half of this chapter.

SUGGESTED READING

For a thorough discussion of the issues of generalizing extremal principles see Santilli (1978, 1983). A survey of issues in scientific explanation is given by Woodward (2009). For a discussion of the relation of intuitive time asymmetry and the asymmetries of statistical physics see Chapter 10 of Sklar (1993). Reichenbach's notion of "screening off" can be found in Chapter IV, Section 22 of Reichenbach (1956).

CHAPTER 22

Retrospective and conclusions

Classical dynamics has a very special place within our theoretical description of the world. For one thing, the theory has had a "lifetime" within physics that is nothing short of astonishing. Beginning with the earliest attempts at a characterization of motion and its causes within ancient Greek science, developing slowly but with some sureness through the Islamic and medieval European eras, exploding into a grand synthesis in the Scientific Revolution, and showing still further important development in its foundations and in applications from then to the present, the theory's place in science is one not of years or centuries but of millennia.

Even now, after having been displaced by relativistic theories and quantum theories and no longer being considered the central "truth" of theoretical physics, the theory still surprises us with new formulations, new applications and new interpretations. Theories in fundamental physics are typically formulated with a characterization of basic states and their possible configurations and changes that is formally similar to the kinematics of dynamics, and with a characterization of changes of these states over time by means analogous to the dynamical parts of classical dynamics. And in the case of relativistic and quantum dynamics the crucial role played by classical dynamics is even clearer since, contrary to some radical "revolutionary" views of the history of science, the important ancestral relationship of the classical dynamical concepts to those of the newer theory is clear.

But classical dynamics also provides us with illuminating lessons to be learned when we look at developed scientific theories from a philosophical perspective. In particular we learn that simple models of methodology that tell us that there is nothing more to theories than hypothesis, empirical test and acceptance or rejection miss out on much that is important about theories.

We learn from classical dynamics and its history that "one and the same" theory can be formulated and reformulated in a variety of ways. And these multiple formulations carry with them distinctly different philosophical perplexities. Consider, for example, the way in which least-action formulations of dynamics carried (and still carry) with them distinct issues about the legitimacy of explanations that are redolent of teleology.

We learn from classical dynamics that a theory can do its work while leaving us in perplexity as to what kind of a world the theory is describing. Even more, we learn that interpretive perplexities can survive through ever deeper understanding of the theory at the formal level and through multiple transformations of our philosophical understanding of theories in general. Despite the vast increase in our knowledge since the days of the Leibniz–Clarke dispute, we are still puzzled by the nature of "absolute" acceleration and its causes. Even transformed by Mach, and enriched by the discovery of special and general relativity and of modern cosmology, the fundamental questions remain enigmatic.

We learn from classical dynamics that it isn't just the fundamental posits of the theory that matter. The specific applied versions of the theory and their consequences can be repositories of hidden interpretive puzzles. Consider, for example, the slow dawning of the realization that there was a world of systems that inhabited a space between that of "simple determinism" and "pure chance." That is, consider the slow development from Poincaré to modern chaos theory of the theory of systems that are deterministic but – in principle – unpredictable. Or the fact that the onset of turbulence still remains a mystery long after the full explication of the Navier–Stokes equations.

We also learn from classical dynamics that ever deeper understandings of what the theory was really trying to tell us about the nature of the world may be available to us only after the theory has been replaced in our corpus of accepted science by its successors. It is only in the light of the results of Minkowski in framing the spacetime picture for special relativity that we come to realize the possibility of a Galilean spacetime for Newtonian dynamics, and only after general relativity has geometrized gravity that a curved-spacetime theory for Newtonian gravitation can be imagined. Can one then hope (as Einstein surely did) that it will only be when quantum theory is replaced by some successor that its mysteries will finally be resolved?

References

Albert, D. 2000. *Time and Chance*. Cambridge, MA: Harvard University Press.

Alexander, H., ed. 1956. *The Leibniz–Clarke Correspondence*. Manchester: University of Manchester Press.

Baird, D., ed. 1998. *Heinrich Hertz: Classical Physicist, Modern Philosopher*. Dordrecht: Kluwer.

Barbour, J. 2001. *The Discovery of Dynamics*. Oxford: Oxford University Press.

Barbour, J. and Bertotti, B. 1977. "Gravity and inertia in a Machian framework," *Il Nuovo Cimento B*, **38**, 1–27.

Barbour, J. and Bertotti, B. 1982. "Mach's principle and the structure of dynamical theories," *Proceedings of the Royal Society A*, **382**, 295–306.

Barbour, J. and Pfister, H., eds. 1995. *Mach's Principle: From Newton's Bucket to Quantum Gravity*. Boston, MA: Birkhäuser.

Barrow-Green, J. 1997. *Poincaré and the Three Body Problem*. Princeton, NJ: Princeton University Press.

Blackmore, J., ed. 1972. *Ernst Mach: His Life, Work and Influence*. Berkeley, CA: University of California Press.

Blackmore, J., ed. 1992. *Ernst Mach – A Deeper Look*. Dordrecht: Kluwer

Bradley, J. 1971. *Mach's Philosophy of Science*. London: Athlone Press.

Clagett, M. 1959. *Science of Mechanics in the Middle Ages*. Madison, WI: University of Wisconsin Press.

Descartes, R. 1983. *Principles of Philosophy*. Dordrecht: Reidel.

Devaney, R. 1986. *An Introduction to Chaotic Dynamical Systems*. Menlo Park, CA: Benjamin.

Diacu, F. and Holmes, P. 1996. *Celestial Encounters: The Origins of Chaos and Stability*. Princeton, NJ: Princeton University Press.

Dugas, R. 1988. *A History of Mechanics*. New York: Dover.

Earman, J. 1970. "Who's afraid of absolute space," *Australasian Journal of Philosophy*, **48**(3), 287–319.

Earman, J. 1989. *World Enough and Space-Time*. Cambridge, MA: MIT Press.

Evans, J. 1998. *The History and Practice of Ancient Astronomy*. Oxford: Oxford University Press.

Galileo. 1914. *Dialogues Concerning Two New Sciences*. New York: Dover.

Galileo. 1967. *Dialogue Concerning the Chief World Systems*. Berkeley, CA: University of California Press.

Garber, D. 1992. *Descartes' Metaphysical Physics*. Chicago, IL: Chicago University Press.

Goldstein, H. 1950. *Classical Mechanics*. Cambridge, MA: Addison-Wesley.

Hanca, J., Tulejab, S., and Hancova, M. (2004). "Symmetries and conservation laws," *American Journal of Physics*, **72**(4), 428–435.

Herival, J. 1965. *The Background of Newton's Principia*. Oxford: Oxford University Press.

Hertz, H. 1956. *The Principles of Mechanics Presented in a New Form*. New York: Dover.

Lagrange, J. 1997. *Analytical Mechanics*. Dordrecht: Kluwer.

Lanczos, C. 1970. *The Variational Principles of Mechanics*. New York: Dover.

Lewis, D. 2009. "Ramseyan humility," in D. Braddon-Mitchell and R. Nola (eds.), *Conceptual Analysis and Philosophical Naturalism*. Cambridge, MA: MIT Press.

Kuhn, T. 1957. *The Copernican Revolution*. Cambridge, MA: Harvard University Press.

Mach, E. 1959. *The Analysis of Sensations*. New York: Dover.

Mach, E. 1960. *The Science of Mechanics*. LaSalle, IL: Open Court.

Moser, J. 1973. *Stable and Random Motions in Dynamical Systems*. Princeton, MA: Princeton University Press.

Noll, W. 1974. "The foundations of classical mechanics in the light of recent advances in continuum mechanics," in W. Noll, *Foundations of Mechanics and Thermodynamics*. New York: Springer-Verlag.

Neuenschwander, D. E. 2010. *Emmy Noether's Wonderful Theorem*. Baltimore, MA: Johns Hopkins University Press.

Newton, I. 1947. *Mathematical Principles of Natural Philosophy (The Principia)*. Berkeley, CA: University of California Press.

Pannekoek, A. 1961. *A History of Astronomy*. New York: Dover.

Poincaré, H. 1952. *Science and Hypothesis*. New York: Dover.

Price, H. 1996. *Time's Arrow and Archimedes' Point*. Oxford: Oxford University Press.

Putnam, H. 1962. "What theories are not," in E. Nagel, P. Suppes, and A. Tarski (eds.), *Logic, Method and the Philosophy of Science*. Stanford: Stanford University Press. Reprinted in H. Putnam, *Mathematics, Matter and Method*, vol. I, pp. 215–227, 1975. Cambridge: Cambridge University Press.

Reichenbach, H. 1956. *The Direction of Time*. Berkeley, CA: University of California Press.

Santilli, R. M. 1978. *Foundations of Theoretical Mechanics, I: The Inverse Problem in Newtonian Mechanics*. New York: Springer-Verlag.

Santilli, R. M. 1983. *Foundations of Theoretical Mechanics, II: Birkhoffian Generalizations of Hamiltonian Mechanics*. New York: Springer-Verlag.

Saunders, S. "Hertz's principles," in D. Baird (ed.), *Heinrich Hertz: Classical Physicist, Modern Philosopher*. Dordrecht: Kluwer, pp. 123–154.

Sklar, L. 1974. *Space, Time and Spacetime*. Berkeley, CA: University of California Press.

Sklar, L. 1980. "Semantic analogy," *Philosophical Studies*, **38**, 217–234. Reprinted in L. Sklar, *Philosophy and Spacetime Physics*, 1985. Berkeley, CA: University of California Press.

Sklar, L. 1993. *Physics and Chance*. Cambridge: Cambridge University Press.

Sneed, J. 1971. *The Logical Structure of Mathematical Physics*. New York: Humanities Press.

Smith, P. 1998. *Explaining Chaos*. Cambridge: Cambridge University Press.

Stein, H. 1967. "Newtonian space-time," *Texas Quarterly*, **10**(3), 174–200.
Sternberg, S. 1969. *Celestial Mechanics*. New York: W. A. Benjamin.
Su, S.-C. and Chu, M.-C. 2009. "Is the Universe rotating?," *Astrophysical Journal*, **703**(1), 354–361.
Swerdlow, N. M. and Neugebauer, O. 1984. *Mathematical Astronomy in Copernicus' De Revolutionibus*. New York: Springer.
Szabó, I. 1977. *Geschichte der mechanischen Prinzipien*. Basel: Birkhäuser.
Taub, L. 1993. *Ptolemy's Universe*. Chicago, IL: Open Court.
Truesdell, C. 1968. *Essays in the History of Mechanics*. New York: Springer-Verlag.
Westfall, R. 1971. *Force in Newton's Physics*. London: Macmillan.
Westfall, R. 1980. *Never at Rest: A Biography of Isaac Newton*. Cambridge: Cambridge University Press.
Whittaker, E. T. 1937. *A Treatise on the Analytical Dynamics of Particles and Rigid Bodies, with an Introduction to the Problem of Three Bodies*, 4th edn. Cambridge: Cambridge University Press.
Winter, A. 1947. *The Analytical Foundations of Celestial Mechanics*. Princeton, MA: Princeton University Press.
Wittgenstein, L. 1958. *Philosophical Investigations*. Oxford: Blackwell (especially items 244–309).
Woodward, J. 2009. "Scientific explanation," *Stanford Encyclopedia of Philosophy* (online).

Index

Printed in the United States
by Baker & Taylor Publisher Services

Printed in the United States
by Baker & Taylor Publisher Services